WHAT IS CALCULUS?

CHRIS MCMULLEN, PH.D.

What is calculus?
Learn the basic concepts (without the hard math)
Chris McMullen, Ph.D.

www.improveyourmathfluency.com
www.monkeyphysicsblog.wordpress.com
www.chrismcmullen.com

Cover design by Melissa Stevens
www.theillustratedauthor.net
Write. Create. Illustrate.

Zishka Publishing
ISBN: 978-1-941691-49-6
Mathematics > Calculus

Table of Contents

Introduction

Would you like to learn what calculus is without having to solve hundreds of difficult problems? Regardless of your math background, you can learn what calculus is, why it was developed, and how it is applied. Whether you never made it further than algebra and would like to appreciate the art of calculus, or if you've actually taken calculus but would like to better understand the meaning of limits, derivatives, and integrals, this book will take you on a tour of the main concepts.

The author, Chris McMullen, Ph.D., has over thirty years of experience teaching physics, a subject which involves both mathematics and conceptual understanding. Students of physics need to apply their math skills while also understanding how the math relates to concepts like energy or torque. The author has taught both calculus-based physics to engineering, physics, and math students as well as conceptual physics to students who are not majoring in math or science. In this book, the author uses his experience of helping students understand the main concepts to show students of all levels the wonders of calculus.

1 What is algebra?

Before we address the question of what calculus is, we should first address the question of what algebra is, since calculus is done using the language of algebra. If you've ever had any exposure to algebra, a variety of ideas may come to mind, such as working with variables, factoring expressions, solving equations, the quadratic formula, the slope of a straight line, or rules for exponents. But what exactly is algebra?

In a way, algebra is almost like a foreign language. Consider a student who has progressed through mathematics in elementary school. Students first learn to count. They learn to add, subtract, multiply, and divide whole numbers. They learn how to work with fractions, decimals, and percents. They discover negative numbers, exponents, and radicals. Up to this point, all the arithmetic involves working with numbers like 4, like 357, or like 6.825. Then they open an algebra textbook and see expressions such as $4x + 5$ or $y^2 - 3y + 8$. Suddenly there are letters in the math.

Algebra introduces the concept of a **variable**, which is a letter that is used to represent a number. A student sees an example where the solution shows that the variable x equals 3, but then in another problem it turns out that the same letter x equals 8, and in yet another problem the variable x happens to be 4. In each problem the same symbol x can represent a different number.

Variables, like x, y, or t, provide a more abstract but also a more advanced framework for thinking about mathematics. Algebra allows us to express mathematical ideas purely in terms of symbols. This abstract framework lets us prove mathematical ideas (called lemmas and theorems), like the Pythagorean theorem ($a^2 + b^2 = c^2$), in general terms rather than specific cases with numbers. Using variables to represent mathematical ideas is also quite practical. In physics and engineering, symbolic formulas like the ideal gas law ($PV = NRT$) or Ohm's law ($V = IR$) make it easy to calculate physical quantities like pressure or current. In chemistry, symbols are also used to represent elements in compounds and reactions. For example, $2H_2 + O_2 \rightarrow 2H_2O$ represents a specific chemical reaction (the formation of water). The coefficients of 2 in front of H_2 and H_2O balance the chemical reaction and tell chemists that they need twice as many moles of diatomic hydrogen (H_2) gas as diatomic oxygen gas (O_2) in order to form water molecules (H_2O). Don't sweat these formulas; the main idea here is that the idea from algebra of using symbols proves to be very helpful not only in mathematics, but also in physics, chemistry, and engineering.

The idea of using symbols to express mathematical ideas is one important aspect of algebra. Another important aspect of algebra is that it provides systematic techniques for solving for unknown quantities. For example, algebra has rules for isolating the unknown in a relatively simple equation like $6x - 8 = 2x + 12$, gives you a formula for solving quadratic equations like $x^2 - 3x - 5 = 0$, and offers a few different methods for solving a system of equations like $2x + 5y = 16$ and $3x - 2y = 5$.

First you express a mathematical idea using symbols in the form of one or more equations, and then you apply the rules of algebra to **solve** for the variables (which are represented by symbols). If you want to figure out what value a quantity (like the volume of a gas tank or the time of an airplane's arrival) is needed to satisfy some condition, algebra provides a way to solve for the value of the unknown.

We'll illustrate this with a few examples. Don't get bogged down in the specifics of the math itself. The purpose of these examples is to help you appreciate the underlying ideas, not to focus on the mathematical techniques themselves.

First, let's review a few common terms. These terms are helpful for referring to different parts of equations. It's sometimes helpful to say things like "the coefficient of x." This will make sense if you know what the term coefficient means, but will seem confusing otherwise. We'll start off with a term that we've already mentioned. **Variables** are unknown quantities represented by symbols. The most commonly used variable is x, but any letter like y or t (or even Greek letters) can serve as a variable. The value of x might turn out to be 4 in one problem, but 16 in a different problem. The values of the symbols vary from problem to problem. In contrast, a **constant** has a definite numerical value. For example, 2 and 105 are constants. Where the language gets a little tricky is when a symbol is used to represent a constant. This is common with the Greek letter pi (which equals 3.14159265... with the digits continuing endlessly without ever forming a repeating pattern). Another example with a symbol that represents a constant is the symbol c in physics, which represents the speed of light in vacuum. Since the speed of light in vacuum is always the same value, it's a constant. The distinction between a variable and a constant is whether the value is subject to change or if the value remains the same.

A **coefficient** is a constant that multiplies a variable. For example, in $7x^4$, the coefficient is 7, and in 24y, the coefficient is 24. In the expression $6x + 3y$, the coefficient of x is 6 and the coefficient of y is 3. One tricky rule in algebra is that if you don't see any constants multiplying a variable, the coefficient is one. For example, in $x + 2$, the coefficient of x is one (and the 2 is just a constant; it's not really the coefficient of a variable[1]).

Here is an example that uses all of these terms. In the expression $3x + 5$, the variable is x, which is some unknown number, the coefficient of x equals 3, and 5 is a constant.

An algebraic **expression** doesn't have an equal sign or inequality, whereas an algebraic **equation** includes an equal sign. For example, $3x + 5$ is just an expression, whereas $3x + 5 = 17$ is an equation. You can **simplify** an expression. For example, $2x + 3 + x + 2$ simplifies to $3x + 5$. You can **solve** an equation. For example, it turns out that $x = 4$ solves the equation $3x + 5 = 17$, as we'll see momentarily. **Terms** are separated by + and − signs (and by =, < , or > signs). For example, the equation $3x + 5 = 17$ has three terms: 3x, 5, and 17.

A big part of algebra is trying to solve equations to determine the values of the variables. For example, consider the equation $3x + 5 = 17$. It basically asks, "Which number can you multiply by 3 and then add 5 to get 17?" Algebra teaches strategies for solving for variables. For the equation $3x + 5 = 17$, algebra tells you to subtract 5

---

[1] If you know that $x^0 = 1$, provided that x is nonzero, you could say that 2 is the coefficient of x^0. But why introduce a variable where there isn't one? It can happen in more advanced algebra courses when you are specifically studying general properties of polynomials. It can also happen in calculus. In Chapters 6-7 we'll learn about limits. In calculus, one might work with the expression x^t and investigate the limit as t approaches zero.

from both sides to get $3x = 12$, and then divide by 3 on both sides to get $x = 4$.[2] This process is called **isolating the unknown**. Originally, the unknown x is together with the 3 in $3x$ and this is next to the constant 5. At the end of the solution, $x = 4$ has the variable x isolated all by itself.

Not all equations can be solved by isolating the unknown this same way. Algebra provides different techniques for solving different kinds of problems. For example, one way to solve an equation of the form $x^2 + 5x - 3 = 0$ is to use the **quadratic formula**. Don't freak out; this is a conceptual book, so we're not going to solve the quadratic equation, and the reader isn't expected to be familiar with the quadratic formula.[3] But since this is a conceptual book, we will take a moment to consider the quadratic formula in terms of concepts without doing algebra. This will be helpful because it will show an example of how algebra is applied in the real world.

One example of where the quadratic formula is used is in physics problems. Suppose that a rock is thrown straight upward with an initial speed of 15 m/s. A physics student could use the formula $x = 15t - 4.9t^2$ in order relate the position of the rock (relative to its starting location) to the elapsed time. A problem might ask, "At which times is the rock 7.2 meters above its initial position?" A student would replace x with 7.2 and then manipulate the equation using algebra to put it in the standard form $4.9t^2 - 15t + 7.2 = 0$. Again, don't sweat the actual algebra. Also, if you're thinking, "Oh no! I don't know physics," relax. You don't need to know physics to appreciate this book. If any ideas from physics or other subjects are important in this book, they will be presented in such a way that you wouldn't actually need to be familiar with them beforehand.

Presently, we're examining the quadratic equation $4.9t^2 - 15t + 7.2 = 0$. In algebra, these are all just numbers, but in physics, these numbers have meaning. Just from reading the previous paragraph, you may have already guessed that 15 m/s was the initial speed of the rock and 7.2 meters is the height of the rock relative to its starting point. The other number, 4.9 meters per second squared, is one-half the value of gravitational acceleration. For the purpose of this example, it's not really necessary to understand what gravitational acceleration is, but so you don't feel left in the dark,

---

[2] It's easier to check the answer than it is to solve for the answer. Replace x with 4 to see if it solves the given equation. We get $3(4) + 5 = 12 + 5 = 17$, which agrees with $3x + 5 = 17$.

[3] For those who aren't afraid of a little math, the quadratic formula below gives the possible answers for x to a quadratic equation in the standard form $ax^2 + bx + c = 0$.

$$x = \frac{-b \pm \sqrt{b^2 - 4ac}}{2a}$$

4.9 is one-half of 9.8 meters per second squared, which is the value of earth's gravity near its surface. This means that an object that is thrown straight upward or downward near earth's surface gains or loses 9.8 m/s of its speed each second. Our rock started moving 15 m/s upward, so it will lose 9.8 m/s of this speed each second on the way up. After 1 second, subtract 9.8 from 15 to determine that it will then be moving 5.2 m/s. After approximately another half-second, the rock will run out of speed, after which it will start gaining 9.8 m/s of speed per second as it falls back towards the ground.

The main idea here is that the rock travels upward and then falls back to the ground, and the equation $4.9\,t^2 - 15\,t + 7.2 = 0$ can help us figure out how much time elapses before the rock is 7.2 meters above its starting position. Without doing any algebra, and without knowing anything whatsoever about physics, it's possible to come to the following conclusion. There can be two different answers for the time. Take a moment to see if you can figure out how. Remember, you don't need to know anything about algebra or physics. Just knowing that the rock goes up and then comes back down is enough information to figure out that there can be two possible values for t. (But if you don't figure it out before reading the explanation below, don't feel bad. It isn't the math that makes physics hard for some students – although the math is the primary hurdle for many students – but the ability to reason out the connection between the math and the concepts, and that's exactly what is involved here.)

Here is how we can reason out that there are two possible answers for the time. Try to visualize the rock moving up and then falling back down. (This is illustrated below.) As the rock climbs upward, it will eventually be 7.2 meters above its starting position. If the rock still has speed when it is 7.2 meters above its starting point, the rock will continue to climb even higher. Eventually, the rock will run out of speed, which will occur at the apex of the rock's trajectory. The rock will then fall back down towards the ground. On the way down, there will again be a moment when the rock is 7.2 meters above its starting location. So, there are two possible answers for the time. The smaller value of time occurs as the rock climbs upward, and the larger value of time occurs as the rock falls back downward.

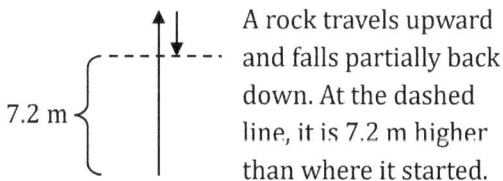

7.2 m

A rock travels upward and falls partially back down. At the dashed line, it is 7.2 m higher than where it started.

A student who is fluent in algebra could use the quadratic formula to determine that the two answers for time in this example are t = 0.596 seconds and t = 2.465 seconds. [4] These algebraic solutions agree with our conceptual reasoning. The rock first reaches the point 7.2 meters above its starting point in less than one second (more precisely, at t = 0.596 seconds), continues to climb higher, and after falling back down it reaches the point 7.2 meters above its starting point for the second time after two seconds (more precisely, at t = 2.465 seconds).

The quadratic equation illustrates an important lesson from algebra. Sometimes a problem has more than one solution. In our example, the quadratic formula provided two different possible values for the variable t. In this case, there wasn't a single unique answer. Some problems do have a single unique answer, but other problems have multiple possible answers. It's even possible for a problem to have no solution. For example, suppose that you throw a rock straight upward with an initial speed of 15 m/s and ask the question, "When will the rock be 1 kilometer above its starting position?" In this case, if you proceed to carry out the calculation (using 1000 meters for the final position, since one kilometer is equivalent to 1000 meters), if you do this correctly, you would need to take the square root of a negative number. Your calculator would give you an error message since the square root of a negative number isn't real (because any number squared will be nonnegative; if you square a negative number, for example, that's positive). In this case, the math is telling you something important. There isn't a real solution because the rock will never be 1000 meters above its starting position if you throw it a mere 15 m/s. [5]

Algebra also involves other ideas, like factoring expressions, rules for working with exponents, properties (like the commutative, associative, and distributive properties),

---

[4] If you try this on your own and would like to check these answers, you can compare your answer with the solution below. Compare $4.9t^2 - 15t + 7.2 = 0$ with $at^2 + bt + c = 0$ (using t instead of x, since in this example the variable is time) to identify a = 4.9, b = −15, and c = 7.2. Now use the quadratic formula (see the previous footnote).

$$t = \frac{-(-15) \pm \sqrt{(-15)^2 - 4(4.9)(7.2)}}{2(4.9)} = \frac{15 \pm \sqrt{225 - 141.12}}{9.8}$$

$$t = \frac{15 \pm \sqrt{83.88}}{9.8} \approx \frac{15 \pm 9.16}{9.8}$$

$$t \approx \frac{15 + 9.16}{9.8} = \frac{24.16}{9.8} \approx 2.465 \quad \text{or} \quad t \approx \frac{15 - 9.16}{9.8} = \frac{5.84}{9.8} \approx 0.596$$

[5] Of course, the formula used assumes that the rock travels only under the influence of gravity. The equation doesn't allow for other situations, such as an eagle catching the rock and soaring up into the clouds.

the equation for a straight line (y = mx + b), solving a system of equations, etc. There isn't just one way to define or explain what algebra is. But so that you have some general sense of what algebra is, two significant points are that **algebra** helps us to express mathematical ideas **symbolically** (that is, in terms of variables) and **algebra** provides systematic methods for **solving** different kinds of equations.

Quick Check (Ch. 1)

Before you move on, see if you can answer these quick questions. If not, it might be worth quickly reviewing Chapter 1 until you can.

1. What is a variable?

2. In the equation $7x - 4 = 10$, what exactly does the term coefficient refer to?

2 What is calculus?

This chapter will help to provide a general sense of what calculus is, while the remaining chapters will discuss specific ideas from calculus (like derivatives, integrals, or optimization) and their applications. One way to get a sense of what calculus can do is by considering what algebra can do versus what it can't do. Recall from Chapter 1 that algebra allows us to express mathematical ideas in symbols (that is, equations with variables) and also provides methods for solving those equations. These aspects of algebra are practical because we can use them to model a variety of real-world problems and make predictions. But as we'll see, there are other real-world problems where algebra alone isn't enough.

   We will consider six examples of how algebra can be used to model real-world problems and make predictions, and then we'll consider how these same examples could be modified so that algebra alone wouldn't be enough. This will help give you a sense of what calculus can do that algebra can't do, and then we'll give a general explanation of what calculus is.

   As our first example, suppose that a ball rolls with constant speed and a child chases the ball with a different constant speed (where the child runs faster than the ball rolls). Given the initial positions of the ball and child along with the speed of the ball and the speed of the child, algebra can be used to predict where the child will be when the child reaches the ball. This is a relatively simple problem (once you have numbers to work with), but we'll see later how this problem can be modified such that calculus is needed to solve it.

For our second example, consider a ramp that has a flat surface. Given the height and length of the ramp, algebra tells us how to find the slope of the ramp: rise over run. The slope tells us how steep the ramp is. If an object is placed on the ramp, given the coefficient of friction between the object and the ramp, physics could be applied to determine whether or not the force of friction would be enough to prevent the object from sliding down the ramp.[6]

Our third example occurs in a parking lot at night. The pavement is flat and horizontal. There is one light at the top of a post and a truck parked some distance away. Given the height of the lightbulb, the dimensions of the truck, and the distance between the post and the truck, algebra (along with trigonometry[7]) could be used to determine the length of the truck's shadow.

In our fourth example, a box slides across horizontal ground in the absence of air resistance. We don't know how the box got into motion in the first place, and it doesn't matter. Somehow, the box is moving. Nothing is currently pushing the box to make it move.[8] Friction between the box and ground will decrease the speed of the box until it eventually comes to rest. Given the initial position of the box, the initial speed of the box, and the coefficient of friction between the box and the ground, algebra can be used to predict the speed of the box at some later time. Algebra can also be used to predict how far the box will travel before it comes to rest.[9]

As our fifth example, suppose that a light ball and a heavy ball are glued to the opposite ends of a rod. Given the mass and radius of each ball and the mass and length of the rod, algebra can be used to find the balancing point of the system. In physics,

---

[6] Physics students can do this by drawing a free-body diagram and applying Newton's second law. What's important here at the moment isn't the physics, but that this problem can be solved using algebra (perhaps with a little trigonometry); calculus isn't needed to solve this problem.

[7] Trigonometry is the study of right triangles. You don't really need to use trigonometry for this problem though, as long as you know about similar triangles in geometry.

[8] The box has inertia, which is the natural tendency of any object to keep moving once it is already in motion. That's why we wear seatbelts in automobiles. If a car suddenly comes to a halt, the passengers would fly forward because they have inertia, but the seatbelt (if used) restrains them in such a situation.

[9] Students learn how to do this in a physics course. They would begin by drawing a free-body diagram and then applying Newton's second law to determine the acceleration. Then they would use the equations of one-dimensional uniform acceleration. It's not the physics that matters here though; our current focus is on whether or not this problem can be solved using algebra, and this problem can.

the balancing point is called the center of mass, and there is an algebraic formula for finding it.

In our last example, there are two points, A and B, where point A is higher than point B. A straight wire joins points A and B. When a bead is placed along the wire at point A, the pull of gravity causes the bead to slide down to point B. Given the angle of the wire, the distance from point A to point B, and the coefficient of friction between the bead and the wire, if earth's gravity is assumed to be constant, algebra (along with a little trigonometry) can be used to predict when the bead will reach point B.[10]

Now we'll consider how to modify each of these examples so that algebra alone wouldn't be enough. But calculus could be used to solve the problem.

In the first example, a child chases a ball. What if the child doesn't run with constant speed? What if the child's speed changes, and even the child's acceleration is non-uniform? Algebra is good when speed or acceleration are constant, but calculus is needed to solve the general case of non-uniform acceleration. (In case you may be wondering what acceleration is, it's the rate at which velocity changes. Also, velocity is a combination of speed and direction. Speed tells you how fast something moves, whereas velocity tells you both how fast it moves and which way it's heading.)

In our second example, an object is placed on a ramp, and the issue is whether the object would stay in place or slide down the ramp. If the ramp is flat, algebra tells us the slope of the ramp from its height and length. But what if the ramp is curved? Given the equation of the curve, we need to apply calculus to determine how steep the ramp is at a particular point. The slope changes along the surface of a curved ramp.

Our third example has a light shining in a parking lot at night, and we were considering the length of a truck's shadow if the truck is parked. But what if the truck begins driving and we wish to know the speed with which the shadow gets longer? When the length of the shadow changes, this becomes a calculus problem. Are you seeing a pattern yet? These examples have something in common. (If you don't see it yet, don't worry. We'll discuss it after finishing the examples.)

In the fourth example, a box is sliding along horizontal ground in the absence of air resistance. But what if there is air? In the more realistic case that there is air, the force of air resistance is proportional to some power of the speed of the box. When the box moves faster, the force of air resistance is stronger, but as the box loses speed,

---

[10] Here is another physics problem that can be solved using Newton's second law. Why? Most people have real-life experience with motion (from walking, throwing balls, etc.), and Newton's second law governs motion, so problems that involve Newton's second law tend to be somewhat relatable.

the force of air resistance diminishes. Since the force of air resistance changes, it takes calculus to solve this problem.

Our fifth example has two balls glued to the opposite ends of a rod, and we were thinking about the balancing point (or center of mass) of the system. What if we wish to find the balancing point of a triangle, a semicircle, or some other general shape? A uniform sphere or uniform rod has its balancing point at its center, but once we start considering general shapes like triangles, semicircles, or cones, calculus is needed to find the center of mass. The nature of this example is somewhat different from the first four examples, but this modification also requires calculus.

In our last example, points A and B (where A is higher than B) are connected by a straight wire. But what if the wire isn't straight? Or what if A is so much higher than B that gravity is significantly weaker at point A than at point B? (This would be the case, for example, if point A lies at the top of a mountain and B lies at the bottom.) In these cases, calculus is needed. Or what if we change the nature of the question as follows. Suppose we ask which shape the wire should have so that the bead reaches point B in the least amount of time? Again, calculus is needed. This is called the brachistochrone problem, and we'll see it in Chapter 20.

Did you notice anything that most of these examples have in common when we modified them so that algebra alone wouldn't be enough? In Examples 1-4 and Example 6, calculus is needed when something is changing. This is a key distinction between calculus and algebra.

One might say in a general sense that **calculus** is the mathematics that describes how things **change**. If slope is constant, algebra suffices, but if the slope changes (such as along a curved ramp), then we need calculus. If speed is constant, or even if acceleration (see Chapter 9) is constant, algebra is good enough, but if the motion is more generally non-uniform (meaning that the velocity and acceleration are both changing), the solution involves calculus. If we wish to determine the length of a shadow, algebra and trigonometry (or algebra along with similar triangles from geometry) work, but if the angle of the rays of light is changing (which causes the length of the shadow to change), it becomes a calculus problem. When the forces (such as gravity and friction) acting on an object are constant, the motion can be modeled and predicted using algebra, but for the general case where one or more of the forces changes (which is the case with air resistance or springs, for example), calculus becomes necessary. In these cases, we are working with rates or slopes and those rates or slopes are changing. This is the general sense in which calculus describes how things change, but this is only one aspect of calculus. (For those students who know a little calculus and who are reading this book in order to try to understand the concepts better, these cases involve derivatives. For everyone else, don't worry if you have no idea what a derivative is. We'll get there in Chapter 8.)

The fifth example, where we wish to find the center of mass of a system, illustrates another important aspect of calculus. If there are just uniform rods and balls, finding the center of mass of the system (or the moment of inertia of the system) is straightforward. In that case, algebra can be used. For most other shapes, like triangles, semicircles, or cones, center of mass isn't so simple. If you place three balls at the corners of a triangle and three rods along the edges of the triangle, that problem is easy, but if you cut a piece of wood into the shape of a solid triangle, finding its center of mass isn't as simple.[11] Calculus can be used to find the center of mass of general shapes, whereas algebra is limited to special cases.

We can appreciate this aspect of calculus by dividing the shape up into several tiny pieces. Suppose, for example, that we have a solid uniform semicircle made of wood. (It's one-half of a wooden disc.) Imagine dividing this wooden semicircle into several little chunks, as illustrated below. We can estimate the center of each chunk, then we could use the algebraic formula for the center of mass using the mass of each chunk along with the approximate location of its center, and this would give us a good estimate of the center of mass of the solid semicircle. It wouldn't be an exact value because we don't know the exact location of each chunk's center. If we make the chunks even smaller, so that the chunks are even more numerous, our approximation for the center of mass of the solid semicircle would become even better. This is basically what calculus does. Calculus takes the **limit** that these chunks become **infinitesimal** (which means as close to zero as you can get yet still be nonzero) in size, and the procedure is referred to as a **Riemann sum**. (We'll consider limits in Chapter 6 and Riemann sums in Chapter 14.) Calculus actually offers a way to do this calculation without actually adding an infinite number of chunks together. (Students who have studied calculus and who are reading this book to better understand the concepts may recognize that this problem is basically an integral. If not, don't worry. We'll explore the concept of an integral in Chapter 14.) This center of mass example may seem abstract and confusing at this point, and if so, that's okay. We'll explore a similar idea (of what an integral is)

---

[11] This particular example can actually be solved without using calculus if your geometry skills are really good. If you know that the three medians of a triangle are concurrent (meaning that they all intersect at a single point) at a point called the centroid and if you know that the centroid cuts each median into a two-to-one ratio, then you know where the balancing point of this wooden triangle is (assuming that the wood is uniform).

in later chapters. For now, the idea is just to see that calculus doesn't only involve change in the form of slopes or rates. It can also involve **area** (Chapters 14 and 16) or center of mass.

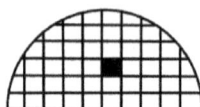

If we want to summarize what calculus is in a nutshell, one way may be as follows. **Calculus** describes how quantities (or functions – see Chapter 5) **change**,[12] such as through **slopes** or **rates**, and it can do so even if those rates or slopes are themselves changing. Calculus also tells you what happens to one quantity when another quantity approaches a particular **limit**. For example, velocity and acceleration are defined as limits where the time interval approaches zero. Calculus also helps you to find the **area** or **volume** enclosed in a general shape (or the length of arc along a general curve), and it helps to calculate other quantities like center of mass, moment of inertia, or electric field where the calculation is similar to finding area or volume (or arc length). These three aspects of calculus are referred to as derivatives, limits, and integrals. (There are also other important topics, such as infinite series.) Students learn about limits first because derivatives are defined in terms of limits, and students learn about derivatives before they learn about integrals because integrals can be found basically by doing derivatives backwards. By the time you finish reading this book, you should have a basic conceptual idea of what limits, derivatives, and integrals mean (even if you can't actually do that math), and you should be familiar with a variety of their applications.

---

[12] More precisely, calculus describes smooth, continuous changes. (What kind of change isn't smooth or continuous? A sudden jump between values, skipping the values in between, isn't continuous. For example, if a quantity equals 20 and suddenly jumps to 25 without ever being anything in between, this sudden jump isn't continuous. If a change isn't smooth, there is a similar sudden change in the rate at which it changes, or its slope. A sudden, abrupt change often isn't smooth. An example that is neither smooth nor continuous is the number of cups of lemonade that a child sells each day at a lemonade stand in the summer; every day that the number of cups sold changes, there is a discontinuous jump, like going from 18 cups sold to 32 cups sold; this lemonade problem is said to be a discrete problem, rather than a continuous problem.) Following is an example of a change that is smooth and continuous. If the pressure applied to the gas pedal of a car changes smoothly, the speed of the car will change smoothly and continuously. If you plot the speed with respect to time, you will get a smooth, continuous curve. Calculus can be used to find the slopes of tangent lines or the area under a smooth, continuous curve.

Fill in the Blank (Ch. 2)

Can you fill in the blank in the sentence below?

1. Calculus can be described as the mathematics of _____.

3 Why was calculus developed?

Isaac Newton and Gottfried Wilhelm Leibniz each independently developed calculus in the late 1600's. Each had realized the value of introducing an **infinitesimal** element, which is now referred to as a **differential** element. The term **infinitesimal** basically means as small as possible without being zero. In this chapter, we will discuss two different kinds of problems which are fundamental to calculus, which involve infinitesimal (or differential) elements. Isaac Newton not only developed calculus, but also developed physics. The first problem that we will consider is fundamental to both calculus and physics.

First, we will consider how to measure or calculate the speed of an object. This was an important problem back in the 1600's. Back then, they didn't have computers, calculators, motion sensors, or even stopwatches. But following the Renaissance that occurred during the 1400's and 1500's, the scientific method had been developed (with important contributions from Copernicus, Francis Bacon, and Galileo) and there were devices for measuring distance and time (for example, Galileo had used pendulums and water clocks).

Now imagine that a rock is traveling through the air, a wagon is rolling along the ground, or a meteor is falling toward the ground and you wish to measure the object's speed. You have instruments like rulers and tape measures for measuring distance and you have instruments like pendulums and water clocks for measuring durations of time. You wish to measure the speed of the object, which indicates how fast the object is moving.

The first difficulty is that the object doesn't have just a single speed to measure. In general, the speed of the object changes, so that every moment it has a different speed. Furthermore, the speed of the object may change non-uniformly. If the object is traveling 20 m/s (meters per second) initially and 40 m/s later, there is no guarantee that the object will be traveling 30 m/s at the halfway point. The object could travel 20 m/s most of the time and rapidly speed up to 40 m/s at the end, or the object could rapidly speed up to 50 m/s in the beginning and slow down to 40 m/s at the end. The possibilities are endless. We want to be able to measure or calculate the speed of the object at any point (not necessarily the halfway point) along the path of its motion.

The easiest kind of speed to measure is the **average speed**. The average speed for the entire trip isn't a very precise measure for the reasons just noted. But it's a good starting point and we will be able to develop a better way of measuring or calculating speed by starting with a definition of average speed. Although it's 'easy' to measure average speed, the way it's done is counterintuitive to most students. We'll compare what's intuitive with what's better to try to understand why average speed is defined the way it is, and eventually we'll see how this relates to calculus.

Most students intuitively want to add the initial speed to the final speed and divide by 2. Why? Because they are told in school that the way to find the average of two numbers is to add them up and divide by 2. For example, the average value of 4 and 12 is 8. When you add 4 and 12 you get 16, then when you divide 16 by 2 you get 8. This definition serves as a good measure of an average in many cases, but in the case of average speed, this definition is lacking, as we will see. It turns out that there are several different kinds of averages. When you add up all the values and divide by the number of values, you are finding the arithmetic mean. Other measures of an average include a weighted average, the median, the geometric average, and there are others. We'll explore why the arithmetic mean doesn't quite work for average speed, and then we'll discuss what works better. (It turns out that a weighted average works better, but you won't need to know anything about weighted averages to follow along.)

Suppose a toy car is moving 4 m/s initially and is moving 12 m/s in the final position. We'll try to convince you that 8 m/s will not, in general, be a good measure of the average speed. We'll use three examples to try to convince you of this. Most students aren't yet convinced after the first example, but after the third example you might finally be ready to give in. It isn't easy. When your brain has been trained to add 4 and 12 to get 16 and then divide by 2 to get 8, it's not easy to accept that another method you haven't ever heard about may be better. One of the tenants of the scientific method is to strive to be open-minded and let measurement-based analysis aid your reasoning.

In the first example, suppose that the toy car travels exactly 4 m/s in a straight line on horizontal ground until it travels 300 meters, at which point it reverses direction and travels exactly 12 m/s until reaching its starting point. Can you think of any reason why the average speed should be **less than** 8 m/s? There is a good reason that the average speed should be less than 8 m/s, and a few students are able to figure this out on their own. Most students don't think of this on their own though, so if you need to read on for the explanation, you shouldn't feel embarrassed; that's normal.

The trick is to think about time. Does the toy car spend more time traveling 4 m/s, or does the toy car spend more time traveling 12 m/s? The answer is that the car spends more time traveling at the slower speed (4 m/s). This should make sense. If you want to drive home from work, you should know that if you drive slower, it will take more time.[13] Students who know the rate equation for constant speed know that distance equals rate times time, which means that time equals distance divided by rate. You could use this equation to determine that the first part of the trip takes 75 seconds and the second part of the trip takes 25 seconds. Since the car spends most of its time traveling 4 m/s and only a quarter of its time traveling 12 m/s, the average speed should be less than 8 m/s. If we find a weighted average where the average is weighted by time, we would indeed find an average where the speed is less than 8 m/s.

We don't need to use a formula for weighted average though. If we define **average speed** as the total distance traveled divided by the total time, we get the same answer that a weighted average would provide. This is the way that physicists define average speed, and we'll see that if we apply it, we get an average speed that is less than 8 m/s. The toy car travels 300 meters forward and backward, so the total distance traveled is 600 meters. The car spends 25 seconds going forward and 75 seconds coming backward, so the total time is 100 seconds. Divide the total distance traveled (600 meters) by the total time (100 seconds) to find that the average speed is 6 m/s (which, as advertised, is less than 8 m/s). Most students aren't quite convinced at this stage, so we'll look at two more examples that are more dramatic.

Suppose that the toy car starts out with one of its wheels on crooked so that it's only traveling 4 m/s. After one minute, we pick up the toy car and fix the wheel, after which time it travels 12 m/s. It spends 999 minutes traveling 12 m/s. Now think about this. The toy car spent 1 minute traveling 4 m/s and it spent 999 minutes traveling 12 m/s. The toy car spent 99.9% of its time traveling 12 m/s, so how could it make sense in this case to add 4 to 12 and divide by 2 to declare that its average speed is 8 m/s? If instead you take the total distance traveled and divide by the total time, you'll find that the average speed is 11.992 m/s.[14] This should make more sense.

---

[13] If you drive faster, you'll probably arrive home in less time, provided that you don't get in an accident and don't get pulled over for speeding. Let's assume that the toy car isn't involved in any accidents.

[14] For those who aren't afraid of a little math. One minute equates to 60 seconds. Multiply 4 m/s by 60 seconds to find that the car travels 240 meters at 4 m/s. Now multiply 999 by 60 to find that the second part of the trip takes 59,940 seconds, and multiply this by 12 m/s to find that the car travels 719,280 meters at 12 m/s. The total distance is 240 + 719,280 = 719,520 meters and the total time is 60 + 59,940 = 60,000 seconds. The average speed is 719,520 / 60,000 = 11.992 m/s. We're not doing calculus here. This is just arithmetic, using the rate

If you spend almost all of your time moving 12 m/s and almost none of your time moving 4 m/s, your average speed should be pretty close to 12 m/s, right? Some students still aren't convinced by this, so we'll do one more example, and if that doesn't convince you, perhaps nothing will.

In our last example, suppose that you get in a car which is parked at your home, and which is initially at rest. You turn the car on. When you first begin driving, your car is moving very slowly down the driveway. Once the car is in the street, you gradually press on the accelerator to get up to speed. You slow down and temporarily come to rest at stop signs and signals, and accelerate to reach higher speeds in between. Eventually, you arrive at your workplace, at which point you slow down and park your car.

Now think about this trip. What was the car's initial speed? Zero! It was parked at your home. What was your car's final speed? Also zero! It's now parked at your workplace. If you add your initial speed to the final speed and divide by two, you would get zero. Does it seem reasonable to say that your average speed for the trip was **zero**? That's what you get if you add the initial speed to the final speed and divide by two. You might have spent most of your time traveling 30 m/s or 50 m/s, for example. Zero should seem quite unreasonable as an average speed in this case. If you divide the distance traveled by the total time, you'll get a much more reasonable value for the average speed. Hopefully, you're now convinced that dividing the total distance traveled by the total time is a better measure of the average speed than adding the initial and final speeds and then dividing by two.

Average speed is found by dividing the total distance traveled by the total time, but what we would really like to do is measure or calculate the speed of an object at any particular moment (or position) along its path. When a police officer pulls someone over for speeding, the police officer doesn't say, "Your average speed for the whole trip was 50 mph and you were driving in a 35-mph zone." The police officer might say, "You were traveling 50 mph when you passed the fire truck." Of course, a police officer can use radar[15] to measure your speed precisely at any given moment. For now, we'd like to think of how to measure or calculate your speed like a radar gun can, but without having that much technology at our disposal. There weren't any radar guns available in Isaac Newton's time.

---

equation from algebra. That rate equation is handy when speed is constant, but calculus will help is figure out what to do when speed changes.

[15] In case you may be wondering, a radar gun uses the Doppler effect to measure speed. By sending out a series of waves and examining how the frequency of reflected waves is shifted compared to the transmitted wave, it is possible to determine a car's speed.

Suppose that an object is traveling along a street. The object could be a car, a bird, or a piece of gravel. It's not important what the object is. What is important is that the speed of the object is likely to be changing, so that the object has different speeds at different times. Let's pick one point on the street and think about determining how fast the object is moving there. This point could be a crosswalk, for example. Our question is, "How fast is the object moving as it passes the crosswalk?" This is referred to as an **instantaneous speed**. The instantaneous speed is the speed of the object at a particular instant, in contrast to the average speed which is averaged over a longer duration.

Let point A be the initial position of the object, point P be on the crosswalk, and point Z be the final position of the object. Note that P lies somewhere between A and Z, but might not be the halfway point. If we measure the distance from A to Z and the total travel time, we could find the average speed of the object, but the average speed could conceivably be quite different from the instantaneous speed of the object as it passes through point P (on the crosswalk).

Here is an example with numbers. Suppose that the object is a car. The car begins from rest at point A, gains speed until it reaches a maximum speed of 60 m/s, then slows down, coming to rest at point Z. The distance from A to Z is 500 meters and the trip lasts 20 seconds. The average speed is 500 meters divided by 20 seconds, which equals 25 m/s. We don't know where point P is. We only know that point P lies between points A and Z. The speed of the car at point P could be anywhere between zero and 60 m/s. The average speed for the entire trip is 25 m/s. In this example, if point P happens to be when the car travels 60 m/s, the instantaneous speed would be more than double the average speed. The point here is that the average speed for the whole trip isn't a good measure of the instantaneous speed.

What would be better? If we want the average speed to be a better measure of the instantaneous speed at point P, we should pick two points closer together. The problem in our example is that A and Z are 500 meters apart. This gives a lot of time for the speed of the object to change. If points A and Z were closer together, there would be less opportunity for the speed of the object to change considerably, and the average speed would be a better measure of the instantaneous speed.

So let's introduce two new points, B and Y. Place point B halfway between A and P, and place point Y halfway between P and Z, as shown below. Measure the distance from B to Y and the time it takes for the object to travel from B to Y, and use this information to determine the average speed from B to Y. This should be a better measure of the instantaneous speed at point P (than when we found the average speed from A to Z). In our example, points B and Y are separated by 250 meters. If the car takes 8 seconds (for example) to travel from B to Y, the average speed from B to Y is 31.25 m/s. Since the trip from B to Y only lasts 8 seconds, there isn't as much time for the car's speed to change (as when it travels 20 seconds from A to Z).

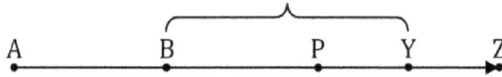

We can do even better if we use two points that are closer together than B and Y. Suppose point C lies halfway between B and P, and point X lies halfway between P and Y, as shown below. If we find the average speed from C to X, that will be a more precise measure of the instantaneous speed at point P (than the average speed from B to Y or the average speed from A to Z).

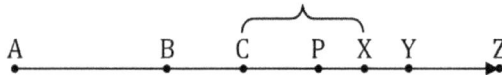

If we keep choosing points that are closer to point P, we will continue to get a better value for the instantaneous speed at point P, provided that our measuring devices are sophisticated enough to make the measurements precisely. If we choose two points that are just a few centimeters apart, it wouldn't suffice to use Galileo's water clocks or pendulums. In modern times, we could connect a photogate to a computer to measure short time intervals precisely.[16]

Newton and Leibniz imagined letting the distance between the two points become so small that the distance is **infinitesimal**. We call this infinitesimal distance a **differential** element. It's as small as possible without being zero. The corresponding time interval is also infinitesimal. But the ratio of the distance to the time, which is the instantaneous speed, is finite. If we make a graph of distance and time, the instantaneous speed at any given point turns out to be the slope of the tangent line. We will explore this idea in Chapter 8. If the graph has a shape like a parabola or any other curve where there is an algebraic formula for how to make the graph, calculus

---

[16] Most physics labs have tracks, cars, and photogate sensors. A transparent plastic insert with dark horizontal bars attaches to the top of the car. The horizontal bars trigger the photogate as the object passes through it. The computer measures the time that it takes for the bar to pass through the photogate. Knowing the length of the horizontal bar then allows for the speed of the car to be determined.

tells us exactly what the slope equals. For example, for the simple parabola $y = x^2$, students who have learned calculus know[17] that the slope equals $2x$ for any value of x.

Calculus works out the mathematics of calculating instantaneous speeds by using limits, or more generally of calculating the slopes of tangent lines using limits. We refer to this as **differential calculus**, and we will explore its main ideas in Chapters 6-8. The spirit of differential calculus is using shorter and shorter distances (so short that they become infinitesimal) to find the instantaneous value of a rate or a slope.

Thus far in this chapter, we've been thinking about how to measure or calculate the speed of a moving object. Now we will invert this problem as follows. Suppose that an object travels forward, but this time we already know the instantaneous speed at every point along the way, and we would like to use that information to determine the total distance that the object travels. For example, we might have a toy car that measures how fast its wheels rotate and uses this information to display the instantaneous speed on a display at the top of the car.[18] If we make a video of the car, we can use the display to see how fast the car is moving at any given moment. The toy car might start out traveling 6.3 m/s uphill, then as the hill becomes less steep it might be moving 9.4 m/s at the crest, then it might gain speed until it is moving 27.5 m/s at the bottom of the hill, and then its speed might reduce to 15.2 m/s after traveling up-hill at the end. These are just 4 of the many instantaneous speeds that it has; its speed gradually changes in between. At any given moment we can look at the display to see the instantaneous speed. After the car has been traveling 2.4 seconds, the display might read 7.0 m/s, and after the car has been traveling 14.6 seconds, the display might read 22.2 m/s, for example. And this is just one possible trip out of an infinite number of possibilities, just to have one case with a few numbers to help serve as a guide.

We basically have a table of data. We can make a table with the elapsed time in one row and the instantaneous speed in an adjoining row. We can record the time and speed every second, every half second, every 0.1 seconds, every 0.02 seconds, etc. The question is what to do with this data to determine how far the object has traveled.

[17] Students would take a derivative of x^2 with respect to x to find that the slope equals $2x$ in this case. If you don't know about derivatives, don't worry. We'll learn about derivatives in Chapter 8.

[18] One could also count the number of times the wheels have rotated and use the circumference of the wheel to determine the total distance traveled, just like how the odometer of a vehicle tells you how far it has been driven. This would be a good way to check our answer. For now, let's just think about how to calculate the total distance traveled (rather than measure it directly) given the instantaneous speeds.

For each pair of time and speed in the table, we can find the approximate distance traveled for that time interval by multiplying the time interval and speed together. (Why? For a short time interval, the speed is relatively constant. When an object travels with constant speed, the distance traveled equals speed times the time interval.) Then we can add all the distances together to find the total distance traveled. We work out an example below.

First, suppose that we choose to record the time and speed in 2-second intervals starting at 1 second. This gives us the data table below. Each speed is taken at the center of a 2-second time interval. For example, for the interval from 0 to 2 seconds, the speed is roughly 6.7 m/s (measured at 1 second), and for the interval from 2 to 4 seconds, the speed is roughly 7.8 m/s (measured at 3 seconds). Before we multiply speed by time to find distance, we need to be careful here. For example, you don't want to multiply 15.5 m/s times 7 seconds, even though the speed at 7 seconds equals 15.5 m/s. Why not? Because 7 seconds is a moment; it isn't a duration. Each speed applies to a 2-second interval, so for each speed, the corresponding time interval is 2 seconds. We want to multiply each speed by 2 seconds to find the approximate distance traveled for each interval. (Why is it approximate? Because the speed of the car changes. If the car's speed stayed constant from 0 to 2 seconds, multiplying 6.7 by 2 would give us the exact distance traveled during the first 2 seconds, but that isn't the case.)

Time (s)	1	3	5	7	9	11	13	15	17	19
Speed (m/s)	6.7	7.8	12.2	15.5	22.3	26.8	23.9	21.6	18.1	16.3

Multiply 6.7 by 2 to see that the car travels about 13.4 m from 0 to 2 seconds, multiply 7.8 by 2 to see that the car travels about 15.6 m from 2 to 4 seconds, multiply 12.2 by 2 to see that the car travels about 24.4 m from 4 to 6 seconds, and so on. Then we would add the distances 13.4 m plus 15.6 m plus 24.4 m and so on to get approximately the total distance traveled.

We would get a better approximation by choosing shorter time intervals. We could instead measure the car's speed at 0.5 seconds, 1.5 seconds, 2.5 seconds, etc. This gives us the car's approximate speed from 0 to 1 seconds, from 1 to 2 seconds, from 2 to 3 seconds, etc. This would give us a better value for the total distance traveled, since the car's speed wouldn't tend to change as much during a one-second interval as it would during a two-second interval. (We're not going to make a new table, which would be larger than the previous one, and go through all the arithmetic again. A big part of calculus is striving to understand the main ideas. We're doing some arithmetic in this book to try to show you what calculus is, but our goal isn't to do arithmetic; it's to understand ideas.)

If we measure the car's speed at 0.05 seconds, 0.15 seconds, 0.25 seconds, etc., we would get an even better measure of the total distance traveled. The car's speed wouldn't tend to change as much during an interval from 0 to 0.1 seconds as it would during an interval from 0 to 1 second, for example, so each average speed would better represent the speed during the corresponding interval. (If we make the time intervals really short, we'll need high-precision instruments to minimize our measurement error. But let's assume that we can make any desired measurement with negligible error and see where this leads. After all, the theory of calculus doesn't depend on the quality of our measuring devices; that only becomes an issue when we proceed to apply calculus in the laboratory or real world.[19])

Newton and Leibniz imagined letting the time intervals become so short that it is **infinitesimal**. We call this infinitesimal time interval a **differential** element. It's as small as possible without being zero. By working with differential elements, calculus is able to provide an exact answer for the total distance traveled. If we make a graph of speed and time, this method finds the area under the curve. We will explore this idea in Chapter 14. If the graph has a shape like a parabola or any other curve where there is an algebraic formula for how to make the graph, calculus tells us exactly what the area equals. For example, for the simple parabola $y = x^2$, students who have learned calculus know[20] that the expression $x^3/3$ can be used to find the area under the curve.

Calculus works out the mathematics of calculating the total distance traveled by dividing the path up into infinitesimal time intervals, or more generally of calculating the total area under a curve by dividing the path up into infinitesimal intervals. We refer to this as **integral calculus**, and we will explore its main ideas in Chapters 14-17. The spirit of integral calculus is to divide a path up into shorter and shorter intervals (so short that they become infinitesimal) to find the area under a curve.

Isaac Newton developed not only calculus, but also physics, and it isn't a big surprise. It turns out that calculus is inherent in physics. Calculus is needed to formulate

---

[19] It turns out that if we know the speed as a function of time, we can find the distance traveled during any time interval exactly; the measurement issue turns out to be a non-issue in that case. Students who have learned calculus should know that this is done by finding an integral. We'll learn about integrals in Chapter 14.

[20] Students would take the anti-derivative of x^2 with respect to x to get $x^3/3$ (plus a constant). Calculus students know that this is an indefinite integral. If they are also given limits of integration (that is, a definite integral), they would evaluate $x^3/3$ at each limit and subtract to find the area under the curve $y = x^2$. If you don't know about anti-derivatives or integrals, don't worry. We'll learn about anti-derivatives and integrals in Chapter 14. This footnote is just for the benefit of any students who may have already studied calculus.

a proper mathematical description of physics. We have already glimpsed how the basic quantities distance, speed, and time are related by calculus, but that turns out to be just the tip of the iceberg. Formulas for many other quantities like work, power, or electric field also involve calculus. The next chapter will explore some applications of calculus, not only in physics, but also in other branches of science, engineering, and even other subjects like economics.

Quick Check (Ch. 3)

Before you move on, see if you can answer these quick questions. If not, you may wish to review Chapter 3 until you can.

1. What does the term infinitesimal mean?

2. What is a differential element?

4 Where is calculus used?

Most people who ask the question "What is calculus?" want to know more than what calculus is. They often also want to know how calculus is useful. This chapter will take you on a tour of a variety of ways that calculus is used in the real world in science, engineering, economics, and other fields. You're not expected to know physics or any of these other subjects, and you're not expected to be familiar with the calculus yet either – this book is just getting started. But this chapter should give you some appreciation for the variety of ways that calculus is useful.

Calculus has applications whenever there is change, especially when the change is non-uniform. In mathematics, the term 'uniform' means constant. In the simple case that things change uniformly, the problem can be solved just using algebra, but in the more general (and common) case that the change is non-uniform, calculus is needed. For example, for an object that travels with uniform acceleration (which describes how velocity changes, where velocity is a combination of speed and direction), meaning that the acceleration is constant, problems relating distance, speed, acceleration, and time can be solved using algebra. However, calculus is needed to solve the more general problem of non-uniform acceleration.

In most branches of physics, calculus is inherent in the equations because most topics in physics involve quantities that change (and often the change is non-uniform). For example, calculus is needed to model the three-body problem consisting of the earth, moon, and sun. The moon orbits the earth while the earth revolves around the sun. The relative positions of the earth, moon, and sun change non-uniformly. In quantum mechanics, calculus is needed to apply Schrödinger's equation (Chapter 21) to solve the similar (yet quite different) problem of an electron orbiting a positively charged nucleus. As another example, calculus is an inherent part of Maxwell's equations (Chapter 21), which describe light, which is an electromagnetic wave. The electric and magnetic fields oscillate, changing with position and time. Calculus is also commonly used in engineering, which applies the concepts and formulas of physics to solve practical problems.

But physics isn't the only subject where quantities change. One particularly important application of calculus lies in the field of medicine. For example, the rate of blood flow may change depending on the shape or obstruction of blood vessels, the calcium content of bones may erode with respect to time, or the volume of a tumor may change. Even basic measurements of heart rate, oxygen levels, and temperature are subject to change. Calculus helps to quantify how such quantities change, and it helps in the analysis of medical charts. Calculus also provides techniques for how to optimize one parameter by changing other parameters. So, for example, to find the optimum dose of a medicine or the optimal time interval at which to administer the doses, calculus tells us how to go about this. The ability to optimize a parameter makes calculus particularly valuable in the fields of medicine, surgery, and medical instrumentation.

Optimization has applications in many other fields, as well. For example, the CEO of a company is generally interested in how to maximize profit. Management has many decisions to make which affect profit and cost, such as how many employees to hire, how many warehouses to have, or which materials to purchase. When it is known how the various parameters relate to the profit, calculus can be used to find the combination of parameters that maximizes the profit.[21] (We'll explore the calculus of optimization in Chapters 10-12.)

It turns out that calculus has widespread applications. Calculus is applied to launch a satellite into orbit, develop and improve technologies, analyze storms and

---

[21] It doesn't have to be money that is maximized, although this is common with most companies. One can imagine a company that strives to maximize the welfare of a community, for example. Whichever quantity you wish to maximize (or minimize), calculus provides the mathematics for how to do it.

seismic activity, and even attempt to predict stock market behavior.[22] In just about any field, you can find examples of where something changes in a smooth and continuous way, for which calculus is relevant. The world is constantly changing.

Sometimes calculus is even helpful when at first it may not seem like anything is changing. For example, suppose that the two ends of a chain are supported at the same height, such that the chain droops as illustrated below. It turns out that we can determine the equation of the shape of the chain by using calculus. Now ask yourself, what is changing? Since the chain is stationary, it might not seem that anything is changing. But if you think about the slope of the chain, you should observe that the slope is changing. If you imagine a bug crawling across the chain, the bug begins with a descent, its descent lessens, the center is momentarily horizontal, from the center it begins an ascent, and the ascent is steepest at the end. It turns out that the curve isn't a parabola, although it may appear similar. The shape of the curve is called a catenary, and the equation for it can be found by using calculus. Catenaries are common, occurring when an object supported only at its ends hangs due to its own weight. Catenaries appear in engineering when cables or chains are supported at their ends, and they even appear in nature (as the branches of a silky spider web, for example).

To help give you an idea of the wide variety of ways that calculus can be applied, consider the examples that follow.

- Launching a rocket to travel to Mars involves solving a (partial) differential equation, which is an equation involving differential elements.
- Earning interest on a savings account with continuously compounded interest

---

[22] Technical analysis (which applies mathematics to analyze and predict stock market behavior) is just one important aspect in buying and selling shares of stocks. So, while the mathematics can tell investors when the stock market may be more vulnerable to a short-term sell-off or a near-term rally, for example, it isn't the only factor that can affect stock prices. Other considerations include fundamentals (like companies that maintain higher profit margins, pay dividends, or do stock buybacks), investor perception (like long-term growth prospects, not just currently but how these prospects are expected to change in the coming quarters and years), and geopolitics (such as the onset of a war or the threat of new tariffs), which are also important and can sometimes override technical analysis.

applies the concept of a limit.[23]

• Data analysis includes finding the slopes of tangent lines, the area under a curve, and extreme values.

• The Maxwell relations of thermodynamics involve partial derivatives (where a derivative applies the idea of finding the slope of a tangent line, or, equivalently, the instantaneous value of a rate).

• The formula for the volume of a cone, hemisphere, an ellipsoid, or other enclosed 3D geometric region applies a triple integral (where an integral applies the idea of finding the area under a curve). Calculus is an instrumental aspect of analytical geometry.

• In number theory, the Basel problem is an infinite series that begins with 1 and then adds 1/4, 1/9, 1/16, 1/25, 1/36, 1/49, 1/64, 1/81, 1/100, etc. Each time, the series grows by $1/n^2$. The series begins 1, 5/4, 49/36, 205/144, and so on. (For example, the fourth term is 1 + 1/4 + 1/9 + 1/16 = 205/144. The fifth term will be 205/144 + 1/25, the sixth term will add 1/36 to this, and so on.) In decimal form, this series looks like 1, 1.25, 1.36111, 1.42361, 1.46361, etc. The calculus of infinite series can be used to show that this series converges to pi-squared divided by 6 (which is approximately equal to 1.645), meaning that the more terms that are added together, the closer the sum gets to one-sixth of pi-squared (that is, the sum approaches $\frac{\pi^2}{6}$).[24]

• In chemistry, calculus is involved in the instantaneous rates with which

---

[23] For those who may be curious about the math, interest earned depends on $(1 + r/n)^m$, where r is the (nominal) annual interest rate, n is the compounding frequency (which would be 1 for annual compounding or 12 for monthly compounding, for example), and t is the number of years. For the case of continuously compounded interest, n grows infinite (such that 1/n becomes infinitesimal). Calculus can be used to show that $(1 + r/n)^n$ approaches e^r in this limit, where e is Euler's number (2.71828...), such that the principal equals the initial investment times e^{rt}.

[24] You might have noticed that this example seems to be discrete, rather than continuous, and you might remember that we defined calculus in general terms as the mathematics of smooth, continuous changes. Going from 1 to 5/4 is a discontinuous jump (skipping all the numbers in between), and going from 5/4 to 49/36 is another such jump. If you noticed such things, you might be wondering where calculus fits into this. One way that calculus is involved in infinite series has to do with the concept of a limit, that is, thinking about the nth term in the limit that n becomes infinite. Another way is that one way to test whether a series is convergent (the sum of an infinite number of terms remains finite) or divergent (the sum becomes infinite, or doesn't converge to a single finite value) is through an integral test.

reactants form products in chemical reactions. (These are called reaction rates, and they relate the concentration of reactants to time.)

• In biology, calculus is involved in the population growth of a species.

• In finance, calculus is used in marginal analysis, where changing one variable (like cost) affects another variable (like revenue).

• The Lagrangian and Hamiltonian formulations of classical mechanics, which help to find the equation of motion of an object or a mechanical system, apply the calculus of variations (which is an area of advanced calculus formulated around the concept of finding the extreme value of a function, like finding the shortest path[25] connecting two points on a particular surface, like a sphere or a saddle). The Lagrangian and Hamiltonian formulations of classical mechanics can be used to solve problems like the motion of a spherical pendulum or the generalized motion of an object that orbits a planet or star.

• In optics, light takes the path of least time according to Fermat's principle.[26] So when light passes from one medium into another medium (such as from water to glass), it refracts (that is, it changes direction) such that the total path takes the least amount of time (between the initial and final points).[27]

• If a piece of wood is shaped like a solid cardioid (which is a mathematical curve that is heart-shaped), the balancing point of the piece of wood can be found by applying calculus. This is done by performing a double integral (where an integral applies the idea of finding the area under a curve).

• The volume and thickness of geological layers below earth's surface can be determined through integrals.

• In medicine, calculus is used to analyze images from MRI's and CT scans. The computer algorithms that help to enhance the images, reduce noise, interpolate

---

[25] Such a path is called a geodesic. In a plane, the shortest path is a straight line, on a sphere the shortest path connecting any two points is a circular arc, and for a general surface, the calculus of variations provides a method for figuring out exactly what the path is. The calculus of variations is used to solve the brachistochrone problem (Chapter 20).

[26] This is not to be confused with Fermat's mathematical theorems, such as Fermat's Last Theorem, which states that $a^n + b^n = c^n$ doesn't have any solutions where a, b, and c are all positive integers when n is 3 or higher. (When n = 2, there are integer solutions. The special case n = 2 corresponds to the Pythagorean theorem. For example, $3^2 + 4^2 = 5^2$. When n is 3 or higher, it turns out that there aren't any integer solutions.)

[27] Light travels faster in water than it does in most types of glass, so to minimize the total time, the path will have more distance in water and less distance in glass. We'll see an example of this in Chapter 20.

and extrapolate data, and identify biomarkers apply techniques from calculus.
• When a substance undergoes a phase change (like water does when ice melts into liquid water or when liquid water boils to form steam), the Clausius-Clapeyron equation helps to determine the equation of the coexistence curve. A plot of pressure and temperature for a substance (referred to as a PT diagram) reveals three distinct regions, one for each of the three (primary) phases of matter. For water, it is solid (ice) at low temperature and high pressure, gaseous (steam) at high temperature and low pressure, and liquid when pressure and temperature are both sufficiently high.[28] The border between any two of these three phases is called the coexistence curve, and there can be one special point on the coexistence curve where all three phases coexist (as in the triple point of water, which occurs at 273 Kelvin and 618 Pascal). The equations for the coexistence curves involve calculus.

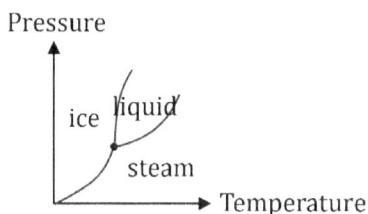

• In probability and statistics, calculus is used (often, with numerical approximations) to calculate probabilities, expected values, and standard deviations involving continuous probability distributions, such as the Gaussian distribution (or normal distribution), which is a bell-shaped curve. When independent measurements are repeated a large number of times, if the measurement error depends on random chance, they tend to follow a Gaussian distribution.
• The theory of black holes in astrophysics involves calculus.
• If you open a packet of sugar and pour it into coffee or tea, the rate at which it dissolves involves calculus. When you order a donut (which has the shape of a single-holed ring torus), its volume is found using calculus. Thus, you even find calculus in coffee and donuts!

---

[28] Here, 'high' is a relative term. We know that water has liquid form at room temperature and standard air pressure, for example, and any temperature above zero degrees Celsius, or 32° Fahrenheit, at standard air pressure. How is this a high temperature? Well, the lowest possible temperature, zero Kelvin, corresponds to negative 273 Celsius; room temperature is extremely high compared to zero Kelvin.

Pop Quiz (Ch. 4)

After reading Chapter 4, which applications do you remember? Take a moment to test your memory.

5 What is a function?

Calculus is taught using functions, so before we begin the calculus, we'll take a moment to try to develop a good understanding of exactly what a function is. We'll begin with the definition of a function, but most students understand what a function is better by example than from the definition, so after defining what a function is, we'll provide some examples. Then we will consider some important kinds of functions that appear in calculus, like trig functions. Don't worry; you won't need to know any trigonometry. We'll just learn the basics of what a few common trig functions mean, so that we will be able to better appreciate what calculus does in the remaining chapters.

A **function** basically receives input and provides output according to some rule such that it provides exactly one output for each unique input. To many students, this definition seems just as Greek as the term function, so if you're thinking "Huh?" right now, you're not alone. The idea of a function should start to make sense as we consider examples. You might also be wondering, "What are input and output?" Good question! In the context of functions, input and output are just numbers.

We'll work out examples that you can imagine holding in your hand. Suppose that you open a birthday present to find a device with a keypad and a display. The keypad lets you enter a real number. It could be a whole number (like 17 or 83,144), a negative number (like –9), a fraction (like 2/3 or 6-1/2), a decimal (like 0.027 or 128.6), or a root (like the square root of 2 or like the cube root of 5.12).[29] When you type the number and press ENTER, another number appears on the display.

---

[29] However, if you try to enter the square root of a negative number, like $\sqrt{-2}$, you receive an error message. Such a number isn't real. To see this, consider $\sqrt{9}$. This means, which number squared equals 9? There are two possible answers since $3^2 = 9$ and $(-3)^2 = 9$. Now suppose that you want to find $\sqrt{-2}$. This means, which number squared equals –2. No real number squared can equal –2 because any nonzero number squared is always positive. In advanced math, you learn that the square root of a negative number is imaginary. Imagine that!

The number that you enter is called the input. The number that appears on the display is called the output. After playing with the device for a while, you realize that if you enter the same input again, you get the same output as before. For example, every time you enter 4, the output is 17, and every time you enter 2/3, the output is 13/9. Evidently, the device gives you the same output for any given input. This device satisfies the definition of a function, and we'll use it to give some examples.

In the first example, when you enter 4, the output is 17, when you enter 7, the output is 50, when you enter 5, the output is 26, and when you enter –2, the output is 5. What does this function do? Can you predict what the output will be if you enter 10? You might want to consider this for a moment before you read the answer in the next paragraph.

This function does two things. First, this function squares the input, and then the function adds one. The function is $f(x) = x^2 + 1$. The first part of this, $f(x)$, is the notation for a **function**. The f is the symbol for the output and the x is the symbol for the input. The parentheses don't mean to multiply f by x.[30] Rather, in the context of calculus, the parentheses tell you that x is the input and f is the output. We read $f(x)$ as "f of x." The variable in parentheses (in this case, x) is called the **argument** of the function. The second part, $x^2 + 1$, is what this particular function does. It squares the input and then adds one. [Note that x^2, which we read as "x squared," means to multiply x by itself: $x^2 = x$ times x. For example, when $x = 4$, x^2 becomes $4^2 = (4)(4) = 16$. That is, 4^2 is 4 times 4. Also note that $(4)(4)$ means to multiply 4 by 4, whereas $f(x)$ means f is a function of x (but doesn't mean to multiply f by x). Although parentheses are often used to indicate multiplication, when we use variables in the form $f(x)$, $g(t)$, or $y(u)$, for example, the student is expected to remember that this is the notation of a function, and doesn't mean to multiply.]

- When the input is $x = 4$, the output is $f(4) = 4^2 + 1 = 16 + 1 = 17$.
- When the input is $x = 7$, the output is $f(7) = 7^2 + 1 = 49 + 1 = 50$.
- When the input is $x = 5$, the output is $f(5) = 5^2 + 1 = 25 + 1 = 26$.
- When the input is $x = -2$, the output is $f(-2) = (-2)^2 + 1 = 4 + 1 = 5$.
- When the input is $x = 10$, the output is $f(10) = 10^2 + 1 = 100 + 1 = 101$.

---

[30] Parentheses often do represent multiplication. For example, $(3)(4)$ means to multiply 3 by 4 to make 12. As another example, $x(x+2)$ means to multiply x by $(x+2)$. So, it is easy for students to look at the notation for a function, $f(x)$, and incorrectly expect this to mean to multiply f by x. This is one of the many pitfalls that calculus students must navigate during a course. The notation $f(x)$ represents a function of the variable x, and doesn't mean to multiply f and x together.

That was our first example. The function was $x^2 + 1$. What does this function do? It squares a given number and adds one. For example, when the input is $x = 5$, the output is $5^2 + 1 = 26$. To predict what the output will be when the input is 10, replace x with 10 in the formula $x^2 + 1$ to get $10^2 + 1 = 101$. If you enter 10 in our device, it will display 101. This is just one possible function.

As you enter more and more possible inputs, you notice that the output is never smaller than 1. When the input is zero, the output is one, and if the input is nonzero, the output is always greater than one. That's because x^2 can't be negative. You can enter any real number for the input, but the output is always a real number greater than or equal to one. We say that the **domain** of this function is all real numbers (meaning that the input, x, can be any real number), while the **range** of this function is 1 to infinity (meaning that the output is never smaller than 1). The domain and range are important considerations in calculus because they tell us if the input or output is restricted. In this case, you can't get negative output or any value less than one, so the range of this function is restricted.

Now you press the RESET button. The device transforms into some different function. With this new function, when you enter 1, the display reads –1, when you enter 6, the display reads 9, when you enter 8, the display reads 13, and when you enter –3, the display reads –9. Can you figure out what this function does? See if you can predict what the output will be if you enter 5.

This new function doubles the input and subtracts three: $g(x) = 2x - 3$. We're calling this function $g(x)$, so as to distinguish it from the $f(x)$ used in the previous example; g is just the name of the function, like x is the name of the variable. The formula $2x - 3$ says to double the input and then subtract three.

- When the input is $x = 1$, the output is $g(1) = 2(1) - 3 = 2 - 3 = -1$.
- When the input is $x = 6$, the output is $g(6) = 2(6) - 3 = 12 - 3 = 9$.
- When the input is $x = 8$, the output is $g(8) = 2(8) - 3 = 16 - 3 = 13$.
- When the input is $x = -3$, the output is $g(-3) = 2(-3) - 3 = -6 - 3 = -9$.
- When the input is $x = 5$, the output is $g(5) = 2(5) - 3 = 10 - 3 = 7$.

That was our second example. The function was $2x - 3$. This function doubles a given number and then subtracts three. For example, when the input is $x = 6$, the output is $2(6) - 3 = 9$. To predict what the output will be when the input is 5, replace x with 5 in the formula $2x - 3$ to get $2(5) - 3 = 7$. If you enter 5 on our device, it will display 7.

Unlike our first function (where the range was restricted to be from 1 to infinity), neither the domain nor the range are restricted in the function $g(x) = 2x - 3$. Put another way, the domain and range both extend from negative infinity to positive infinity; they are the set of all real numbers. If you put in a very large input, like $x = 1,000,000$, you'll get a very large output, and if you put in a very negative input, like $x = -1,000,000$, you'll get a very negative output, with no limits to the size of the input or output. Contrast this with our first example, where the output could never be less than one.

Press the RESET button again. The function will be different this time. With this third function, when you enter 4, the display reads 2, when you enter 9, the display reads 3, when you enter 36, the display reads 6, and when you enter 8, the display reads 2.828427125. Also, when you enter −5, the display gives you the message, "not a real number." Try to figure out what this function does.

This third function takes the (positive) square root of the given number: $h(x) = \sqrt{x}$ (that is, the square root of x). Students who have learned algebra may know that we can write this function as $h(x) = x^{1/2}$, since an exponent of one-half means to find the square root, but this point is not important. The important point is that $h(x)$ returns the square root of x.

- When the input is $x = 4$, the output is $h(4)$ = square root of 4 = 2.
- When the input is $x = 9$, the output is $h(9)$ = square root of 9 = 3.
- When the input is $x = 36$, the output is $h(36)$ = square root of 36 = 6.
- When the input is $x = 8$, the output is $h(8)$ = square root of 8 = 2.818427125.

(The digits of this irrational number go on forever without ever forming a repeating pattern, but the display only shows nine decimal places.)

That was our third example. The function was $\sqrt{x}$. For example, when the input was 36, the output was the square root of 36, which equals 6 (because $6^2 = 36$).

There is an important point about the square root in this example, which is significant to what a function is (and what a function isn't). To understand this point, first you need to know that, in general, a square root can have both positive and negative roots. For example, consider the equation $y^2 = 4$, which asks the question, which number squared equals 4. The answer is the square root of 4. There are actually two answers. One answer is $y = 2$ because $2^2 = (2)(2) = 4$, but a second answer is $y = -2$ because $(-2)^2 = (-2)(-2) = 4$ also. When negative two is multiplied by itself, the answer is positive four because the two minus signs cancel out. The complete solution to $y^2 = 4$ includes the two possible answers $y = 2$ and $y = -2$.

However, in our third example, we only considered the positive roots (not the negative roots), and for a very good reason. **A function can't be multi-valued.** For each input, x, there can only be one output, $f(x)$. There is an important distinction to be made here. There can be two different inputs that result in the same output, but there can't be two different outputs corresponding to the same input. In our first example, $f(x) = x^2 + 1$, observe that $x = 3$ and $x = -3$ are two different inputs that have the same output: $f(3) = 3^2 + 1 = 10$ and $f(-3) = (-3)^2 + 1 = 10$. It's okay for two inputs to result in the same output. In contrast, in our third example, $h(x) = \sqrt{x}$, we can't include both positive and negative roots because that would result in two outputs for the same input. For example, when $x = 4$, we can't allow $h(4) = 2$ and $h(4) = -2$ because then $h(4)$ would be multi-valued.

For this reason, calculus adopts the convention that a square root always implies the positive root only. There is a little subtlety here that tends to confuse students. When we see an algebraic equation like $x^2 = 25$, where we need to square root both sides of the equation in order to solve for x, we state that the answer is $x = 5$ or $x = -5$ because both possible answers satisfy the given equation. In contrast, if we are working with a function like $f(x)$ = square root of 25, in the context of functions only the positive root is implied. In the context of functions, this convention is important because a function can't be multi-valued.

If a function were multi-valued, this would cause problems in calculus. For example, when we learn about derivatives, we will see that finding a derivative equates to finding the slope of a tangent line. If a function had two different values of f for the same value of x, it would have two different tangent lines at x, and hence two different slopes, which would be a problem. Since a function can't be multi-valued, we don't have to worry about this possibility in calculus. Similarly, when we learn about integrals, we will see that a definite integral corresponds to the area under the curve. If a function had two different values of f for the same value of x, the area under the curve wouldn't make sense. Which of the two points would mark the 'top' of the curve? Since a function can't be multi-valued, we also don't have to worry about this possibility in calculus.

Our third example illustrates another important aspect of functions. The input can't be negative for $h(x) = \sqrt{x}$. If the input, x, were negative, $h(x)$ wouldn't be real because no real number squared can be negative (as discussed in Footnote 29). If you try to find the square root of negative five, for example, a calculator will give you a domain error. The domain of the function $h(x)$ includes all nonnegative numbers (meaning that the input, x, can be 0 or any positive real number). The range of $h(x)$ also includes all nonnegative numbers (meaning that the output is also never negative).

In general, how can you tell whether some expression is a valid function or not? If you make a graph with the function f on the vertical (or y) axis and its variable on the horizontal (or x) axis, it's easy. Just use the **vertical line test**. According to the vertical line test, if it's possible to draw a vertical line that intersects the curve at two different points, the graph isn't a function. Why not? Because it would be multi-valued. It would have two different values of f for the same value of x.

For example, consider the two plots below. The plot on the left isn't the plot of a function because it is possible to draw a vertical line that intersects the curve at two different points. The plot on the right is the plot of a function because it isn't possible to draw a vertical line that intersects the curve at two different points.

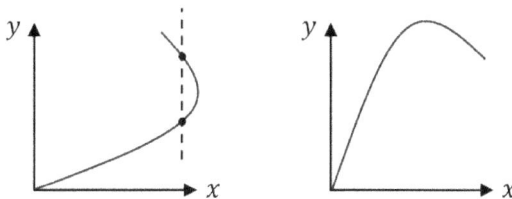

The variable in the argument of the function (that is, the variable that appears in parentheses) is referred to as the **independent variable**, whereas the symbol used to indicate the function is referred to as the **dependent variable** (since the value of the function depends on the value of the independent variable; that is, once you know the value of the independent variable, you can use the mathematical rule to determine the value of the function). For example, with f(x), the independent variable is the argument x and the dependent variable is the function f. As another example, with y(t), the independent variable is the argument t and the dependent variable is the function y.

In summary, a **function** follows some rule and for each value of the independent variable (the input), the function provides exactly one value of the dependent variable (the output). The domain of a function refers to the possible values of the independent variable (the argument that appears in parentheses), whereas the range of the function depends on the possible values of the dependent variable. For example, for the function y(x) = x^4, the rule is x^4. Given a value of the independent variable x, to find the corresponding value of the dependent variable y, the rule says to raise x to the fourth power (which means to multiply x times x times x times x). For example, when x = 3, this rule tells us that y(3) = 3^4 = (3)(3)(3)(3) = 81. When x = 5, this rule gives y(5) = 5^4 = (5)(5)(5)(5) = 625. The domain of this function is that x may be any real number, since any real number can be raised to the fourth power. The range of this function is that y is nonnegative; it can be 0 or any positive real number. Any nonzero number raised to the fourth power is positive. (This is true for even powers, like x^2 or x^6. In contrast, odd powers like x^3 or x^9 can be negative or positive.)

The previous example illustrates why we need the parentheses. We used the parentheses to distinguish the case y(4) from the case y(5). If we simply called each case y, that would be confusing. The notation y(x) refers to the function and its argument in general, whereas y(4) and y(5) are two specific cases (where x happens to be 4 or where x happens to be 5).

Another important kind of function that is used in calculus is the piecewise function. A piecewise function follows different rules over different intervals. The simplest kind of piecewise function is the **step function**. The step function defined below equals zero whenever x is negative and equals one whenever x is greater than or equal to zero. Note how this function follows two different rules, depending on the value of the independent variable. Also, note how this function jumps from 0 to 1 as x turns from negative to zero. This sudden jump is called a **discontinuity**. The discontinuity occurs at x = 0. In calculus, it's important to know if a function has discontinuities because the rules of calculus (namely, derivatives and integrals which we will learn about in later chapters) are valid over regions where functions are continuous.

$$f(x) = 0 \text{ if } x < 0$$
$$f(x) = 1 \text{ otherwise}$$

As another example, consider the function below, which is divided into three pieces. The function below equals x if x is 1 or smaller, equals x^2 if x lies in the range from 1 to 5, and equals $1/x$ if x is 5 or greater. This function has a discontinuity at x = 5, but is continuous at x = 1 (because when x = 1, both x and x^2 equal the same value).

$$g(x) = x \text{ if } x \text{ is 1 or smaller}$$
$$g(x) = x^2 \text{ if } 1 < x < 5$$
$$g(x) = 1/x \text{ if } x \text{ is 5 or greater}$$

The functions that we have considered thus far have been algebraic functions, meaning that we have been able to express the rule using algebraic expressions like $f(x) = 5x^2 - 6x + 2$ or like $h(t) = \sqrt{t}$. In calculus, students work with other kinds of functions besides just algebraic functions. We'll consider a couple of these now.

Trigonometric functions, like sine, cosine, and tangent, are very common in calculus. You don't need to take a course on trigonometry to understand what these functions represent. You just need to be able to draw a right triangle and remember how these functions are defined (see below). A right triangle is a triangle with a **right**

angle, which is where two perpendicular sides meet, forming a 90° angle. For example, in the right triangle shown below, sides a and b are perpendicular. Sides a and b form a right angle.

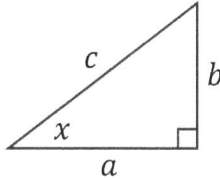

Consider the right triangle illustrated above. The sides are labeled a, b, and c. Observe that two of the sides (a and b, which are called the legs) form the right angle (which measures 90°), whereas the longest side (c, called the **hypotenuse**) doesn't touch the right angle. Consider the angle marked with an x in the triangle above. Of the two legs (a and b), observe that side a is **adjacent** to x, whereas side b is **opposite** to x. Noting which leg is adjacent and which leg is opposite to the angle of interest is important. The three basic trig functions are defined as follows:

- The **sine** function, denoted $\sin(x)$, is the ratio of the opposite side to the hypotenuse. For the triangle shown above, $\sin(x) = b/h$ because b is the opposite side and h is the hypotenuse.
- The **cosine** function, denoted $\cos(x)$, is the ratio of the adjacent side to the hypotenuse. For the triangle shown above, $\cos(x) = a/h$ because a is the adjacent side and h is the hypotenuse.
- The **tangent** function, denoted $\tan(x)$, is the ratio of the opposite side to the adjacent side. For the triangle shown above, $\tan(x) = b/a$ because b is the opposite side and a is the adjacent side.

More generally, these three trig functions are defined as:
$$\sin(x) = \text{opposite/hypotenuse}$$
$$\cos(x) = \text{adjacent/hypotenuse}$$
$$\tan(x) = \text{opposite/adjacent}$$

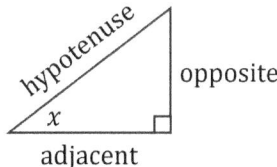

$$\sin x = \frac{\text{opposite}}{\text{hypotenuse}} \qquad \cos x = \frac{\text{adjacent}}{\text{hypotenuse}} \qquad \tan x = \frac{\text{opposite}}{\text{adjacent}}$$

The **tangent** function serves a particularly important role in calculus. Study the triangle shown previously. Observe that side a (which is adjacent to x) is horizontal, side b (which is opposite to x) is vertical, and side c (the hypotenuse) is tilted. What is the slope of side c? The **slope** of a line is a measure of its steepness; the greater the slope, the steeper it appears. In algebra, we learn that the slope of a line equals the **rise** over the **run**. In the diagram above, the rise is the vertical measure, b, while the run is the horizontal measure, a. The slope of side c (the hypotenuse) is equal to b/a (which is the rise divided by the run). If you check the last bullet point above, you'll see that $\tan(x) = $ b/a, which shows that the tangent function gives the **slope** of the hypotenuse (for a right triangle that is drawn like the one above, with horizontal and vertical legs). In Chapter 8, when we explore derivatives, we'll be finding the slopes of tangent lines. The tangent function is aptly named, since the tangent function gives the value of the slope of the tangent line. (If you're not sure what a tangent is, you'll learn about it in Chapter 8.)

As an example of the basic trig functions, consider the right triangle shown below. The legs are the two shortest sides, marked 3 and 4. The sides of length 3 and 4 are perpendicular, forming a right angle. The hypotenuse has a length of 5. The three sides of a right triangle satisfy the **Pythagorean theorem**, which states that the sum of the squares of the lengths of the sides equals the square of the hypotenuse. In this case, 3 squared and 4 squared add up to 25, which agrees with 5 squared. That is, $3^2 + 4^2 = 5^2$. Now consider the angle marked x. The side with length 4 is adjacent to x, whereas the side with length 3 is opposite to x. For this triangle, the basic trig functions are:

- $\sin(x) = $ opp/hyp $= 3/5 = 0.6$.
- $\cos(x) = $ adj/hyp $= 4/5 = 0.8$.
- $\tan(x) = $ opp/adj $= 3/4 = 0.75$.

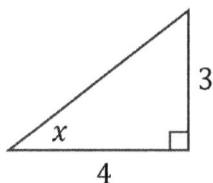

Students who have learned about trigonometry should know that sine squared plus cosine squared equals one. This statement is no different than the Pythagorean theorem. If you try it for the triangle above, you'll see that sine squared plus cosine squared is $0.6^2 + 0.8^2 = 0.36 + 0.64 = 1$. To see how this is the Pythagorean theorem, we'll write it using the fractions: $(3/5)^2 + (4/5)^2 = 9/25 + 16/25 = 25/25$. Just focus on the end of this equation: $9/25 + 16/25 = 25/25$. The three denominators are 25. If you look at the numerators, you'll see that they say $9 + 16 = 25$, which is the same as $3^2 + 4^2 = 5^2$. Look at the previous triangle. This is identical to the Pythagorean theorem.

Two more functions that are common in calculus include logarithms and exponents. Logarithms tend to scare students, at least if they are asked to use them on homework or if they appear on exams. But this is a conceptual book, and just understanding the basic idea of what a logarithm is shouldn't be scary. Logarithms are related to exponents, so what we'll mainly need to know in this book is how logarithms relate to exponents.

In the expression x^p, we call x the **base** and p the **exponent** (or **power**). For example, in 2^4, the base is 2 and the exponent is 4. This means to multiply 2 by 2 by 2 by 2 (with a total of four two's multiplying): $2^4 = (2)(2)(2)(2) = 16$. Exponents are easy. Sometimes the numbers get large and you need a calculator to figure them out, as is the case with 42^{18}. And exponents are convenient. It's much easier to type 42^{18} on your calculator than it would be to repeatedly multiply 42 by itself until you have 18 of them multiplying one another.

Now imagine that a student is doing algebra and that the variable happens to be in the exponent, like $2^p = 512$. This equation says, "Which exponent of 2 makes 2^p equal to 512?" The answer is p = 9. If you try it on your calculator, you'll see that $2^9 = 512$. That problem can be done without a calculator, since the answer happens to be a whole number. Some problems require using a calculator, like $10^y = 73.8$. But wait a minute! How would you figure this out using a calculator? The answer is that you need to use a base-ten logarithm.

We have discovered one reason that logarithms exist. Logarithms let us solve for a variable that appears in an exponent. A **logarithm** function has the form $\log_b(x) = y$, where b is the **base**, the dependent variable y is the exponent, and the independent variable x is equal to b^y. What's important here is that the equation $\log_b(x) = y$ is equivalent to $b^y = x$. So if you see an equation of the form $b^y = x$ and wish to solve for the exponent y, what you need to enter into your calculator is $\log_b(x)$. For example, to consider the equation $10^y = 73.8$. Here, the base is b = 10, x = 73.8, and the exponent is y. To solve for y, enter $\log_{10}(73.8)$ into a calculator. If the calculator has a base-10 logarithm function, you'll just take a base-ten logarithm of 73.8 to get approximately y = 1.868. (Be careful not to use a natural logarithm or a logarithm of a base that isn't 10. For example, if you try this on your calculator and get 4.301, your calculator used a natural logarithm instead of a base-10 logarithm.) It's easy to make a mistake, so it's wise to check the answer. Simply raise the base (b = 10) to the power

1.868 and see if this agrees with 73.8. You should find that $10^{1.868}$ is approximately equal to 73.8, which confirms that y = 1.868 is the answer to this problem.

If you're totally new to logarithms, this probably seems a bit confusing, and that's okay. You shouldn't expect to become an expert on logarithms shortly after learning about them. So let's focus on what really matters for our purposes. The main idea is that the logarithm $\log_{10}(73.8)$ is equivalent to the problem $10^y = 73.8$ and allows us to solve for the exponent y using a calculator. Any logarithm is equivalent to a problem with exponents. If you have an equation where the variable happens to be in the exponent and you wish to solve for the variable, it's possible to do this using a calculator by working with logarithms.

We'll try one more example, this time where every number is a whole number, which will make the math simpler. (If we only used nice round numbers, you might not appreciate that logarithms are ever necessary. Using 73.8 in the previous example illustrates that sometimes you can't solve for the exponent in your head, but that logarithms sometimes serve a useful purpose.)

Consider the problem $10^n = 1,000,000$. This problem is simple enough that if you understand exponents well enough, you can figure out the value of n without using a calculator. Take a moment to try it. The answer is given in the following paragraph.

Which power can you raise 10 to and obtain one million as a result? First note that when you multiply any number by 10, this adds a zero to the number. For example, 23 times 10 equals 230 (see the zero that appeared?), and 1000 times 10 equals 10,000 (which has one more zero than 1000). So how many 10's would you need to have multiplying one another in order to make 6 zeroes? The answer is 6. That is, $(10)(10)(10)(10)(10)(10) = 1,000,000$. We can write this more compactly using exponents: $10^6 = 1,000,000$. You would probably like the number 10^6 if you could put a dollar sign in front of it and if it would be all yours. (But if it were that easy to make a million dollars, it would be worthless.) For the problem $10^n = 1,000,000$, the answer is n = 6. This problem was simple enough that it could be solved without knowing about logarithms. But since our goal presently is to better understand what a logarithm is, let's compare this with the logarithm method. The base is b = 10. To solve for the exponent (n), we need to use $n = \log_{10}(1,000,000)$, which means to find a base-ten logarithm of 1,000,000 on your calculator. If you do this correctly, you'll find that n = 6. (If you get 13.82, you used a natural logarithm instead of a base-10 logarithm.)

What was important this time? The main idea is that the problem $10^n = 1,000,000$ is equivalent to the logarithm $n = \log_{10}(1,000,000)$, which equals 6. When

10 has an exponent of 6, this makes 10^6 = 1,000,000. The logarithm lets us solve for the exponent (which is especially helpful when the answer doesn't turn out to be a whole number).

Logarithms can have different bases. For example, $\log_2(8)$ is a base-2 logarithm (which is equivalent to $2^y = 8$) and $\log_6(5)$ is a base-6 logarithm (which is equivalent to $6^y = 5$). The most common logarithms are base-10 logarithms and natural logarithms. A **natural logarithm** is a logarithm where the base is Euler's number (a mathematical constant with the symbol e, which is approximately equal to 2.71828). The natural logarithm is written as $\ln(x)$ instead of $\log_e(x)$. The base of $\ln(x)$ is implied to be e, which is approximately 2.71828. The natural logarithm $\ln(x) = y$ is equivalent to $e^y = x$. It is called the natural logarithm because it 'naturally' shows up in nature. For example, when radioactive isotopes (like carbon-14 or various isotopes of uranium) decay, it does so with exponential decay of the form e^{-kt} (where t represents time and k is a constant). In order to solve for t, one needs to use a natural logarithm. As another example, simple population growth is modeled as e^{kt}. There are many other cases of logarithmic and exponential behavior in science and engineering, for which the natural logarithm is helpful.

The domain of the logarithm is any positive real number. The logarithm function is undefined when the argument is zero. Why? Because $\log_b(x) = y$ is equivalent to $b^y = x$. If you try to let $x = 0$, this means that you want b^y to be zero. But if b is a positive base, no value of y will make b^y zero.[31] For the same reason, the argument of a logarithm can't be negative. Since the argument can't be zero or negative, the domain includes only positive real numbers. The range of the logarithm function is the set of all real numbers. The logarithm function is negative when the argument is less than the base, and is positive when the argument is greater than the base. For example, $\log_{10}(1000)$ is positive 3 (1000 is greater than 10), whereas $\log_{10}(0.001)$ is negative 3 (since 0.001 is smaller than 10). The first problem equates to $10^y = 1000$, while the second problem equates to $10^y = 0.001$. **When the argument equals the base, the logarithm is one.** For example, $y = \log_{10}(10) = 1$. This problem is equivalent to $10^y = 10$. Which power of 10 equals 1? The answer is the first power: $10^1 = 10$. **When the**

---

[31] The best that you can do is make y a very negative number. If b = 10 and y = −100, then b^y is 10^{-100}, which is very tiny. In algebra, students learn that $w^{-p} = 1/w^p$, such that $10^{-100} = 1/10^{100}$. Since 10^{100} is huge, $1/10^{100}$ is tiny, so 10^{-100} just as tiny. If you make y − −1,000,000, b^y will be even tinier. As y approaches negative infinity (a concept we'll explore in the next chapter), b^y gets closer and closer to zero, without quite ever getting there. That's why the logarithm of zero is undefined; there doesn't exist a finite exponent that would make b^y zero.

argument is one, the logarithm is zero. For example, $y = \log_{10}(1) = 0$. This problem is equivalent to $10^y = 1$. Which power of 10 equals 1? The answer is zero: $10^0 = 1$. Students learn in algebra that any (nonzero) number raised to the power of zero equals one.[32]

If this seems daunting, remember that the main idea is that the logarithm function helps us solve for an exponent. Logarithms are related to exponentials. An exponential has the form e^x or e^{-x} (or more generally e^{kx}). An exponent with base e turns out to be special in calculus (we'll learn in Chapter 8 that it's the one function that equals its own derivative). Other exponentials like 2^x or 10^x have similar behavior to e^x (and can actually be related by logarithms). Sometimes the exponential function e^x is expressed as $\exp(x)$. Writing e^x to represent a function may seem odd, since functions have arguments; it's implied that the exponent is the argument. Writing $\exp(x)$ makes it more obvious that it's a function.

Now you should be a little familiar with a variety of functions. Check your understanding. Do you remember what a function is? You should know two basic properties that all functions have in common. If you've forgotten, you might want to reread a little from earlier in this chapter, so that when we mention functions in later chapters, you have some idea of what a function is.

Quick Check (Ch. 5)

Before you move on, see if you can answer these quick questions. If not, it might be worth quickly reviewing Chapter 5 until you can.

1. What is a function?

2. Given $f(x) = 6x^2 - 9$, what does $f(2)$ equal?

---

[32] This follows from the rule $w^{q-p} = w^q/w^p$. If you let $q = p$, the left-hand side becomes w^0 (since $p - p = 0$) and the right-hand side becomes 1 (since $w^p/w^p = 1$), which shows that $w^0 = 1$ if w is nonzero. There is an exception when w is zero, since w^p/w^p would be zero divided by zero in that case, which is indeterminant. In the context of logarithms, though, the base can't be zero, so this exception isn't relevant.

6 What is a limit?

Calculus begins with the concept of a limit of a function. Recall that we learned the basics of what functions are in the previous chapter. If we have a function $f(x)$, a **limit** indicates what happens to the value of $f(x)$ as the independent variable x approaches a particular value. For example, suppose that $f(x) = x^2 + 4$ and we would like to explore what happens to $f(x)$ as the value of x approaches 3. This particular example turns out to be trivial because if you simply replace x by 3 in the equation, you get $f(3) = 3^2 + 4$ $= 9 + 4 = 13$, which happens to be the correct answer. (Recall that x^2 means to square x, which means to multiply x by itself. When $x = 3$, this becomes $3^2 = 3$ times $3 = 9$.) As we'll see, as x approaches 3, in this example $f(x)$ approaches 13. However, many limits are not so trivial; simply plugging in the desired value of x into the function $f(x)$ doesn't always give the correct answer for the limit, as we'll discover in subsequent examples. Because of this, calculus uses a different method (that is, different from simply plugging the desired value of x into the function) to evaluate a limit. But we'll start out with this trivial example because it's easy to understand that $f(x)$ approaches 13 as x approaches 3 without knowing anything about calculus.

What exactly do we mean by "as x approaches 3" and which "value $f(x)$ approaches"? Good questions. The basic idea of "as x approaches 3" is to consider different values of x that get closer and closer to the value of $x = 3$. For each value of x that we consider, we'll plug that value of x into $f(x)$ and explore what happens to the value of $f(x)$. The pattern that we see will help us predict which "value $f(x)$ approaches" as x approaches 3. We'll explore this numerically in the next paragraphs to help make this clear.

First, there are two different ways that x can approach the value of 3. One way is to consider values of x that are smaller than 3, but which are slowly increasing, getting closer and closer to 3. For example, we might start with 2.5, then 2.7, then 2.9, then 2.95, then 2.97, then 2.99, then 2.999, then 2.9999, etc. These values of x are approaching 3. This is a one-sided limit. In this limit, x approaches 3 from below.[33] Another way is to consider values of x that are greater than 3, but which are slowly decreasing, getting closer and closer to 3. For example, we might start with 3.5, then 3.3, then 3.1, then 3.05, then 3.03, then 3.01, then 3.001, then 3.0001, etc. These values of x also approach 3. This is another kind of one-sided limit. In this limit, x

---

[33] You could alternatively say that such a limit is approached from the left. If you plot $f(x)$ as a function of x, values of x that are less than 3 appear to the left of $x = 3$.

approaches 3 from above.[34] There are two kinds of **one-sided limits**: those that approach a particular value of x from below, and those that approach a particular value of x from above. A one-sided limit tells you in which direction the limit is approached: from below or from above.

Let's evaluate the function $f(x) = x^2 + 4$ when x is approaching 3 from below, starting at $x = 2.5$ as outlined previously.

- When $x = 2.5$, the function equals $f(2.5) = 2.5^2 + 4 = 6.25 + 4 = 10.25$.
- When $x = 2.7$, the function equals $f(2.7) = 2.7^2 + 4 = 7.29 + 4 = 11.29$.
- When $x = 2.9$, the function equals $f(2.9) = 2.9^2 + 4 = 8.41 + 4 = 12.41$.
- When $x = 2.95$, the function equals $f(2.95) = 2.95^2 + 4 = 8.7025 + 4 = 12.7025$.
- When $x = 2.97$, the function equals $f(2.97) = 2.97^2 + 4 = 8.8209 + 4 = 12.8209$.
- When $x = 2.99$, the function equals $f(2.99) = 2.99^2 + 4 = 8.9401 + 4 = 12.9401$.
- When $x = 2.999$, the function equals $f(2.999) = 2.999^2 + 4$ $= 8.994001 + 4 = 12.994001$.
- When $x = 2.9999$, the function equals $f(2.9999) = 2.9999^2 + 4$ $= 8.99940001 + 4 = 12.99940001$.

Examine the pattern of numbers at the ends of the lines above: 10.25, 11.29, 12.41, 12.7025, 12.8209, 12.9401, 12.994001, 12.99940001. As expected, as x gets closer and closer to the value of $x = 3$, the function $f(x)$ gets closer and closer to the value of $f(3) = 13$. If you plug in 2.99999999 for x, it will be even closer to 13. You can imagine a number beginning with 2 which has billions of 9's after the decimal point, and even way more 9's than that. In the limit that the number of 9's following the decimal point becomes infinite, the function becomes virtually identical to 13. That is, as x approaches the value of 3 from below, the function approaches 13. We say that the limit of $f(x)$ as x approaches 3 from below is equal to 13.

Now let's evaluate the function when x is approaching 3 from above, starting at $x = 3.5$ as outlined previously. (We're doing a lot of arithmetic here to try to illustrate a main idea of calculus. In a calculus course, students do a lot of calculus, algebra, and trig, but not so much arithmetic. We're using arithmetic as a tool to help understand concepts of calculus since many people who read this book won't already know trig or precalculus. But remember, the arithmetic isn't our goal; it's just a tool. Our goal is to understand the main ideas.)

---

[34] You could alternatively say that such a limit is approached from the right. If you plot $f(x)$ as a function of x, values of x that are greater than 3 appear to the right of $x = 3$.

- When x = 3.5, the function equals f(3.5) = 3.5^2 + 4 = 12.25 + 4 = 16.25.
- When x = 3.3, the function equals f(3.3) = 3.3^2 + 4 = 10.89 + 4 = 14.89.
- When x = 3.1, the function equals f(3.1) = 3.1^2 + 4 = 9.61 + 4 = 13.61.
- When x = 3.05, the function equals f(3.05) = 3.05^2 + 4 = 9.3025 + 4 = 13.3025.
- When x = 3.03, the function equals f(3.03) = 3.03^2 + 4 = 9.1809 + 4 = 13.1809.
- When x = 3.01, the function equals f(3.01) = 3.01^2 + 4 = 9.0601 + 4 = 13.0601.
- When x = 3.001, the function equals f(3.001) = 3.001^2 + 4 = 9.006001 + 4 = 13.006001.
- When x = 3.0001, the function equals f(3.0001) = 3.0001^2 + 4 = 9.00060001 + 4 = 13.00060001.

Once again, examine the pattern of numbers at the ends of the lines above: 16.25, 14.89, 13.61, 13.3025, 13.1809, 13.0601, 13.006001, 13.00060001. As expected, as x gets closer and closer to the value of x = 3, the function f(x) gets closer and closer to the value of f(3) = 13. If you plug in 3.00000001 for x, it will be even closer to 13. You can imagine a number beginning with 3 which has billions of zeroes after the decimal point followed by a single one, and even way more zeroes than that. In the limit that the number of zeroes following the decimal point (and prior to the first nonzero digit) becomes infinite, the function becomes virtually identical to 13. That is, as x approaches the value of 3 from above, the function approaches 13. We say that the limit of f(x) as x approaches 3 from above is equal to 13.

In this example, it didn't matter whether x approached the value of 3 from above or below. In either case, the limit as x approached 3 of f(x) was equal to 13. That is, whether x approached the value of 3 from above or from below, f(x) approached 13. As we'll see in another example later in this chapter, it's possible for the limits from below and from above to be different. When the limit from above equals the limit from below, we say that the limit exists for the function at that particular value of x. So in this example, the limit as x approaches 3 of f(x) is equal to 13; we don't need to bother saying from above or from below, since both are equal. In this example, the limit exists. The notation for these limits is shown below. The letters "lim" stand for limit, the arrow indicates that x approaches 3, and the limit equals 13. The 3+ means that x approaches 3 from above, while the 3– means that x approaches 3 from below. If the limit from above differs from the limit below, the limit doesn't exist for the function at that particular value of x (we'll see an example of this later).

$$\lim_{x\to3^-}(x^2+4)=13 \qquad \lim_{x\to3^+}(x^2+4)=13 \qquad \lim_{x\to3}(x^2+4)=13$$

Our computations in the previous sets of bullet points, which show that $f(x)$ approaches 13 from above or below as x approaches 3, don't technically 'prove' that the limit as x approaches 3 of f(x) equals 13. Calculus students use delta and epsilon notation to prove limits. We'll wait until the next chapter to discuss what delta and epsilon notation is. For now, we'll explore the concept of a limit informally to try to understand what limits are.

At this stage, it may seem like the entire idea of a limit is unnecessary, but that's mainly because thus far the only limit we have considered has been a trivial case. If you simply replace x with 3 in the equation $f(x) = x^2 + 4$, you get $f(3) = 3^2 + 4 = 9 + 4 = 13$. To help understand how the concept of a limit may prove useful, we'll examine some less trivial cases. As noted earlier in this chapter, it isn't always possible to simply replace x with the desired value in the equation for $f(x)$ to evaluate a limit.

One kind of limit where you can't simply plug in the value of x on your calculator is when x becomes infinite. Your calculator probably doesn't have an infinity button. What you can do is explore what happens to $f(x)$ as x becomes very large. Another thing you can do is make a graph of $f(x)$ and see what happens to the curve as x grows very large.

As an example, consider the function $g(x) = (6x + 5)/(2x - 3)$. This function is a fraction. The numerator is $(6x + 5)$, and it is divided by the denominator, which is $(2x - 3)$. We'll find the limit of g(x) as x becomes infinite. We'll do this informally by calculating $g(x)$ for larger and larger values of x and observing the pattern. (For the limit that x goes to infinity, note that it's only possible to approach the limit from below. You can't approach infinity from above.)

- When $x = 100$, the function equals g(100) = 605/197, which is approximately 3.07107. How did we get this? We replaced x with 100 in the numerator, which is $6x + 5 = 6(100) + 5 = 600 + 5 = 605$, and in the denominator, which is 2(100) − 3 = 200 − 3 = 197. That's where the 605 and 197 came from. Then we divided 605 by 197 on a calculator to get 3.07107. In the subsequent calculations, we'll just show the values of the numerator, denominator, and final value.
- When $x = 1000$, the function equals g(1000) = 6005/1997, which is approximately 3.00701.
- When $x = 10,000$, the function equals g(10,000) = 60,005/19,997, which is approximately 3.00070.
- When $x = 100,000$, the function equals g(100,000) = 600,005/199,997, which is approximately 3.00007.

As we increased x from 100 to $100,000$, $g(x)$ decreased from 3.07107 to 3.00007. Looking at the pattern, you should observe that $g(x)$ is approaching 3 as x grows larger and larger. The limit as x becomes infinite of $g(x)$ is 3 in this example.[35]

It is instructive to compare the functions $p(x) = x$ and $q(x) = 1/x$ as x grows infinite. As you consider larger and larger values of x, the function $p(x) = x$ grows just as large as x (since this function actually equals x). If x is a million, $p(x)$ is also a million. If x is $1,000,000,000,000,000,000$, $p(x)$ is also $1,000,000,000,000,000,000$. As x grows to infinity, it should be clear that $p(x)$ also grows to infinity. We say that the limit of $p(x)$ as x goes to infinity is infinity. But if you think about it, infinity isn't really a limit. Infinity is unlimited; it's the lack of a limit. There is no limit to how large a number can be. We use the word infinity to indicate this. Nevertheless, in a calculus class, if a student is asked what the limit of $p(x) = x$ is in the limit that x goes to infinity, the correct answer is that this 'limit' is infinity.

Let's contrast $p(x) = x$ with the function $q(x) = 1/x$. What happens to $q(x)$ as x grows infinite? You should be able to see that as x grows larger and larger, $q(x)$ becomes smaller and smaller (but still positive). For example, when $x = 100$, we get $q(100) = 1/100 = 0.01$, when $x = 10,000$, we get $q(10,000) = 1/10,000 = 0.0001$, and when $x = 1,000,000$, we get $q(1,000,000) = 0.000001$. If you're familiar with fractions, you may recognize that $q(x) = 1/x$ is the **reciprocal** of x. In the fraction $1/x$, the numerator is always 1 and the denominator is x. As x grows larger, the numerator remains the same (equal to 1) while the denominator grows larger. When the denominator gets bigger and the numerator remains the same, the fraction gets smaller. Try it. When $x = 1,000,000,000$, we get $q(1,000,000,000) = 1/1,000,000,000 = 0.000000001$; here, x is very large, while $1/x$ is tiny. As x grows infinite, $1/x$ becomes **infinitesimal** (meaning as tiny a number as possible while still being positive). In this example, $q(x)$ approaches **zero** in the limit that x goes to infinity. (It approaches zero without ever quite getting there. Just as x will never actually reach infinity, $1/x$ will remain

---

[35] In the ratio $(6x + 5)/(2x - 3)$, calculus students should know that only the terms of the highest power of the polynomials in the numerator and denominator matter in the limit that x goes to infinity. In this case, the highest power in the numerator is $6x$ and the highest power in the denominator is $2x$. The highest power of x is proportional to x (as opposed to x^2 or x^3) in both cases. When the highest power in the numerator and denominator is the same, the limit as x goes to infinity is a nonzero, finite constant. To find this constant, divide the coefficients of the highest powers in the numerator and denominator. These coefficients are the 6 of $6x$ and the 2 of $2x$. When we divide them, we get $6/2 = 3$.

infinitesimal, not quite getting all the way to zero, but you can get $1/x$ as close to zero as you want just by using a larger value of x.)

Another common case where you can't simply plug the value of x into a function is when the function itself grows to infinity (or negative infinity). Even if x is finite, the function may go to infinity. For example, consider the function $f(x) = 1/(x-2)$ as x approaches 2. For this function, we subtract 2 from the value of x, and divide 1 by that. For example, when $x = 6$, we get $f(6) = 1/(6-2) = 1/4 = 0.25$. Let's explore what happens to $f(x)$ as x approaches the value of 2.

When x is larger than 2 and is getting smaller (that is, as x approaches 2 from above), observe that the denominator $(x-2)$ gets smaller (while remaining positive). When the denominator gets smaller, the fraction $1/(x-2)$ gets larger. For example, when $x = 2.1$, we get $f(2.1) = 1/(2.1-2) = 1/0.1 = 10$. When $x = 2.01$, we get $f(2.01) = 1/(2.01-2) = 1/0.01 = 100$. When $x = 2.001$, we get $f(2.001) = 1/(2.001-2) = 1/0.001 = 1000$. The closer x gets to 2 (while still being larger than 2), the smaller the denominator gets and the larger $f(x) = 1/(x-2)$ becomes. As x approaches 2 from above, the quantity $(x-2)$ becomes infinitesimal, while $f(x) = 1/(x-2)$ becomes infinite.

Now consider what happens as x approaches 2 from below. Since x is smaller than 2 in this case, the quantity $x-2$ is negative. (For example, $1.9-2$ equals -0.1, which is negative.) As x gets closer to 2 from below, the denominator $(x-2)$ is a negative number that gets closer to zero, and $f(x) = 1/(x-2)$ is a very negative number. For example, when $x = 1.9$, we get $f(1.9) = 1/(1.9-2) = 1/(-0.1) = -10$. When $x = 1.99$, we get $f(1.99) = 1/(1.99-2) = 1/(-0.01) = -100$. When $x = 1.999$, we get $f(1.999) = 1/(1.999-2) = 1/(-0.001) = -1000$. The closer x gets to 2 (while still being smaller than 2), the closer the denominator gets to zero (while still being negative) and the more negative $f(x) = 1/(x-2)$ becomes. As x approaches 2 from below, the quantity $(x-2)$ becomes infinitesimal (while remaining negative), while $f(x) = 1/(x-2)$ goes to negative infinity (that is, it becomes a very negative number).

In this example, the limit as x approaches 2 of $f(x)$ **does not exist**. Why not? Because the limit of $f(x)$ as x approaches 2 from above is positive infinity, whereas the limit of $f(x)$ as x approaches 2 from below is negative infinity. Since one limit is positive and the other is negative, the two limits (from above and below) aren't equal. When the limit from above differs from the limit from below, the limit does not exist at the particular value of x. We'll see another example of a limit that doesn't exist next, this time where x and the function are both finite.

$$g(x) = 0 \text{ if } x < 0$$
$$g(x) = 1 \text{ otherwise}$$

Consider the step function above, which we discussed in the previous chapter. If we want to find the limit of g(x) as x approaches any negative value, it will be zero since g(x) = 0 for all negative values of x. If we want to find the limit of g(x) as x approaches any positive value, it will be 1 since g(x) = 1 for all positive values of x. But what about the special value x = 0? That's different. Let's consider the limit of g(x) as x approaches zero. If x approaches zero from below, the limit of g(x) is zero. If x approaches zero from above, the limit of g(x) is one. Since these two limits differ (one is zero while the other is one), the limit as x approaches zero of g(x) does not exist.

$$h(x) = 0 \text{ if } x \text{ is } 0 \text{ or negative}$$
$$h(x) = x \text{ if } x \text{ is between } 0 \text{ and } 1$$
$$h(x) = 1 \text{ if } x \text{ is } 1 \text{ or greater}$$

It's possible for a limit to exist where the intervals of a piecewise[36] function meet. To see this, consider the piecewise function above. For example, h(x) approaches zero as x approaches zero from below or above. When approaching from above, we need to use the middle interval (that is, when x is between zero and one). In this interval, h(x) = x. As x approaches zero from above, h(x) = x such that h(x) and x both approach zero. Since the limit as x approaches zero from above or below equals the same value (zero), the limit of h(x) as x approaches zero exists and equals zero. Similarly, you should be able to see that the limit of h(x) as x approaches one also exists and equals one.

$$g(x) = x \text{ if } x \text{ is negative}$$
$$g(x) = 2 \text{ if } x \text{ equals } 0$$
$$g(x) = x^3 \text{ if } x \text{ is positive}$$

---

[36] Do step functions and piecewise functions seem unnatural to you? They actually occur in nature. For example, if a sphere has a uniform distribution of positive charge throughout its volume and you wish to calculate the electric field, it turns out that the electric field has one equation for the region inside of the sphere and another equation for the region outside of the sphere. The electric field in this case is represented by a piecewise function. As another example, in quantum mechanics, consider a problem where two regions of space are separated. A particle has enough energy to be in either of the two regions, but doesn't have enough energy to be in the space between the two regions. Without quantum mechanics, the particle wouldn't be able to travel from one region to the other because it lacks the energy to get there. Yet, in quantum mechanics, a phenomenon referred to as tunneling can occur, where the particle can actually travel from one region to the other. When we solve Schrödinger's equation to find the wave function, the equation is different in the regions and the space between. In this case, the wave function is a piecewise function.

Even more dramatically, it's possible for the limit to exist, yet for the limit to differ from the value of the function at the desired point. This is the case for the piecewise function g(x) above. Whether x approaches 0 from below or above, g(x) approaches zero. Approaching from below, g(x) = x such that g(x) and x both approach zero. Approaching from above, g(x) = x^3. Since x approaches zero, x^3 also approaches zero. Since the limits from below and above are equal, the limit as x approaches zero of g(x) exists and it equals zero (since the limit from below and above are both equal to zero). Although the limit is zero, the function actually equals 2 when x = 0. (This is a peculiar piecewise function in the sense that the middle region only applies to the single point x = 0.)

Now let's look at a more practical limit. Recall that we briefly learned about the basic trig functions (sine, cosine, and tangent) at the end of the previous chapter. When the acute angle of a right triangle is greater than zero but less than 90 degrees, it's straightforward to find the values of sine, cosine, and tangent. We did an example in the previous chapter with a 3-4-5 right triangle.

But what if you want to find the sine, cosine, or tangent of zero degrees or ninety degrees? It's impossible to draw a right triangle with an angle of 0 degrees, or to draw a triangle with two 90-degree angles. (Since the sum of the three angles must be 180, if there were two 90-degree angles, the third angle would need to be zero.) Since it's not possible to draw such a right triangle, how can you find the sine, cosine, or tangent of an angle of zero or ninety degrees? It turns out that it is possible if we apply the concept of a limit.

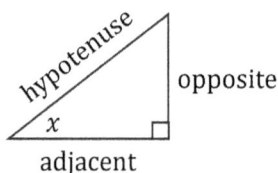

$$\sin x = \frac{\text{opposite}}{\text{hypotenuse}} \qquad \cos x = \frac{\text{adjacent}}{\text{hypotenuse}} \qquad \tan x = \frac{\text{opposite}}{\text{adjacent}}$$

$$\sin(x) = \text{opposite/hypotenuse}$$
$$\cos(x) = \text{adjacent/hypotenuse}$$
$$\tan(x) = \text{opposite/adjacent}$$

Recall the definitions of sine, cosine, and tangent from the previous chapter. In the right triangle above, x is one of the acute angles. The definitions of the sine, cosine, and tangent functions for angle x are given above.

Now imagine that the hypotenuse has a length of exactly one unit. For this special case, $\sin(x)$ will equal the opposite side and $\cos(x)$ will equal the adjacent side.[37] To see what happens as x approaches zero, consider a right triangle where x is very small (yet where the hypotenuse is still one unit in length), such as the right triangle shown below on the left. In the right triangle below on the left, where x is very small, the opposite side is very small, but the hypotenuse is one. As x approaches zero, the opposite side approaches zero. Since the hypotenuse is one unit long, $\sin(x)$ approaches zero (since the sine function is the opposite over the hypotenuse). This is why $\sin(0) = 0$. If you enter $\sin(0)$ on your calculator, you will see that it equals zero. As x approaches zero, the adjacent side gets longer, approaching the length of the hypotenuse. In the limit that x approaches zero, the ratio of the adjacent side to the hypotenuse thus approaches one. This is why $\cos(0) = 1$ (because the cosine function is the adjacent over the hypotenuse). If you enter $\cos(0)$ on your calculator, you will see that it equals one. As x approaches zero, the ratio of the opposite side to the adjacent side approaches zero, since the opposite side approaches zero and the adjacent side approaches one. This is why $\tan(0) = 0$ (because the tangent function is the opposite over the adjacent).[38]

Now consider what happens as x approaches 90 degrees. Consider the right triangle on the right side of the above diagram (where the hypotenuse still has a length of one unit). In the right triangle above on the right side, where x is nearly 90 degrees, the adjacent side is very small, while the opposite side is nearly as large as the hypotenuse. This time, $\sin(90°) = 1$ (because the sine function is the opposite over the hypotenuse, and both are nearly equal) and $\cos(90°) = 0$ (because the cosine function

---

[37] Students who have taken trigonometry should recognize that this corresponds to the unit circle. When we draw a right triangle on the unit circle, the hypotenuse of the triangle and the radius of the circle both have a length of exactly one unit. Some trig concepts are simpler on the unit circle, since the hypotenuse equals one unit.

[38] Students of trigonometry know that tangent equals sine over cosine. So, another way to find $\tan(0)$ is to divide $\sin(0)$ by $\cos(0)$ to get $0/1 = 0$.

is the adjacent over the hypotenuse, and the adjacent approaches zero). The tangent function is interesting in this limit. Since the tangent function is the opposite over the adjacent, and since the adjacent approaches zero, the tangent function grows infinite as x approaches 90° from below. The proper answer is that tan(90°) is undefined.[39]

We'll see in Chapter 8 that the concept of a limit can help us find the slope of the tangent line of a curve or the instantaneous rate of change of a quantity. This practical use of limits has wide applications.

Quick Check (Ch. 6)

Before you move on, see if you can answer these quick questions. If not, you may wish to review Chapter 6 until you can.

1. What is a limit?

2. Given $f(x) = x^2 - 3$, what is the limit of $f(x)$ as x approaches 2? (Maybe you can test this out using a calculator.)

3. Given $g(x) = 5/x$, does the limit of $g(x)$ as x approaches 0 exist? Explain.

7 What are delta and epsilon?

Calculus students learn how to use delta and epsilon notation to evaluate limits. This is an abstract way of thinking that challenges many students. We will focus on the main ideas to help you try to appreciate what many calculus students are taught to do when they solve problems involving limits.

Chapters 6 and 7 both cover limits. Chapter 6 was the easy conceptual chapter, while Chapter 7 is a bit more abstract and technical. Chapter 7 may be the most difficult chapter of this book, so if you struggle with it, that's not totally unexpected. But if you don't fully grasp Chapter 7, you don't need to give up. If you understand the main ideas from Chapter 6, for the purpose of this book, that will be good enough

---

[39] If you're tempted to answer that tan(90°) is infinity, one problem with that is that if you approach tan(90°) from above, you get negative infinity, and so the limit of tan(x) as x approaches 90 degrees does not exist. Students who know trigonometry should know that the tangent function is negative in Quadrant II, where x is greater than 90 degrees and less than 180 degrees. Since tan(90°) approaches positive infinity from below and negative infinity from above, the limit does not exist.

to move on. This pair of chapters make an important point. Most of this book attempts to explain the main ideas of calculus in a friendly, informal way, using a minimum of mathematics. Chapter 7 helps to convey that calculus is much more challenging to students enrolled in a course than this book may make it seem. When you reach the end of this chapter, you should note that the only math involved includes arithmetic and algebra; it's not really fancy or advanced math. What makes this chapter tough for many students (including those who are taking a calculus course) is that the thinking is more abstract, the language is more technical (it's possible for language to be highly technical even without using big or obscure words), and the logic is more formal.[40] The more mathematics courses students take, the more abstract, technical, and formal it becomes. So this chapter will give you a better taste of what a calculus course is like. But remember, the other chapters should seem easier in comparison, and you don't really need this chapter to continue on with the book.

If this seems Greek to you, it should; delta and epsilon are literally the names of letters from the Greek alphabet. To make it friendlier, in this book we will write out the words delta and epsilon instead of using the Greek letters. It may not seem as frightening that way, and you won't have to remember which words (delta or epsilon) go with which Greek symbols. For those who want to know what these Greek letters look like, δ is lowercase delta and ϵ is lowercase epsilon. Note that these are lowercase letters. If you're familiar with the names of various fraternities and sororities, you're probably accustomed to the uppercase letters, which obviously look different from the lowercase letters.

Before we introduce what delta and epsilon mean, we'll get an example started. This will help make delta and epsilon seem more tangible. Consider the function $f(x)$ $= x^2 - 1$. We wish to explore the limit of $f(x)$ as x approaches 4. If we use the informal method from Chapter 6, we would evaluate $f(x)$ at values like 3.9, 3.95, 3.99, and 3.999 to find the limit of $f(x)$ as x approaches 4 from below, and evaluate $f(x)$ at values like 4.1, 4.05, 4.01, and 4.001 to find the limit of $f(x)$ as x approaches 4 from above. If you use this informal method,[41] you'll discover that $f(x)$ approaches $f(4) = 4^2 - 1 = 16 - 1$ $= 15$ from below or from above, such that the limit as x approaches 4 of $f(x)$ is 15.

---

[40] Another factor that makes calculus courses difficult for many students is fluency with algebra and trigonometry. It's amazing how often a student is stuck on a calculus problem not because the calculus is difficult, but because the student has forgotten some important algebra rule or trig identity.
[41] This is a trivial case like the first example of the previous chapter, where if you just plug $x = 4$ into the equation for $f(x)$, you obtain the answer for the limit. As noted in the previous chapter, not all limits can be found by plugging the desired value of x into $f(x)$. Also, the informal method from the previous chapter doesn't constitute a formal proof of the limit. In the current chapter, we'll explore how calculus students use delta and epsilon to construct more formal proofs.

What matters so far in our example are these three points: $f(x)$ is equal to $x^2 - 1$, x approaches the value of 4, and $f(x)$ approaches the value L = 15. We're using L for the value that $f(x)$ approaches. If the limit turns out to exist (which it does, in this example), we'll refer to L as the limit. That is, as x approaches 4, the limit of $f(x)$ is L = 15.

First, we'll introduce the concept of delta.[42] **Delta** is a small deviation from the value that x approaches. In our example, where x approaches 4, delta is a small deviation from 4. The idea is that delta is a small number compared to 4. So in our example, delta might be 0.1, it might be 0.05, or it might be 0.02. Unfortunately, delta and epsilon don't have fixed values. We need to have a range of possibilities in mind. Let's consider a specific value of delta to begin with, in order to help understand what delta and epsilon are, and how they are related. For now, we'll choose delta to be 0.1. Since delta is a small deviation from 4, in our example this means that we're considering values of x from 3.9 to 4.1 (since $4 - 0.1 = 3.9$ and $4 + 0.1 = 4.1$). In this limit example, where x approaches 4, when we set delta equal to 0.1, this means that we're currently thinking about values of x as small as 3.9 or as large as 4.1, and which are approaching 4.

Now let's see how values of x from 3.9 to 4.1 affect the function $f(x) = x^2 - 1$. To do this, we'll evaluate $f(x)$ at the two extreme values: 3.9 and 4.1. When $x = 3.9$, we get $f(3.9) = 3.9^2 - 1 = 15.21 - 1 = 14.21$, and when $x = 4.1$, we get $f(4.1) = 4.1^2 - 1 = 16.81 - 1 = 15.81$. In this case, if x lies between 3.9 and 4.1, the function $f(x)$ lies[43] in the range 14.21 to 15.81. **Epsilon** is a small deviation from the limit L (if it exists) that $f(x)$ approaches. Recall that in this example, L = 15. We'll use our range for the function (14.21 to 15.81) and L = 15 to find a value for epsilon that corresponds to our value of delta. If we subtract $15.81 - 15 = 0.81$, we get 0.81, and if we subtract $15 - 14.21$, we get 0.79. We want the larger difference for epsilon, so epsilon = 0.81 in this case. Why the larger value? If delta is 0.1, such that x is between 3.9 and 4.1,

---

[42] This is backwards from how the delta and epsilon procedure tends to be worded in calculus. Calculus textbooks generally consider a value of epsilon, and then proceed to try to find a corresponding value of delta. We're first introducing delta because it will be easier to understand the connection between delta and epsilon this way. We'll eventually get to the usual calculus wording, where we first consider a value of epsilon and proceed to think about delta. For those who already know that much calculus, just bear with us. We'll get there eventually.

[43] There is more to it than this. Just because $f(3.9) = 14.21$ and $f(4.1) = 15.81$, this doesn't in general guarantee that $f(x)$ lies between 14.21 and 15.81 if x lies between 3.9 and 4.1. For this particular function, that's the case, but it isn't the case for all functions. One way to know is to graph the function (on a graphing calculator or computer, for example).

we want to know how far f(x) may differ from 15, and in the worst-case scenario, that's 0.81. In this example, if delta is 0.1, then a corresponding value for epsilon is 0.81. But we could choose 0.9 or 1 for epsilon, and it would still work.

We really want to do this in reverse. We want to choose a value for epsilon and then consider what the corresponding value of delta would be. That's the opposite of what we did in the previous paragraphs. We'll try this again, but this time we'll pick epsilon first.

Suppose that we wish to choose epsilon to be 0.5. What would the corresponding value of delta need to be? This time, we want f(x) to lie in the range 14.5 to 15.5 (since $15 - 0.5 = 14.5$ and $15 + 0.5 = 15.5$). Which value for delta would ensure that f(x) lies in this range? We obviously need a smaller value of delta than last time because when delta is 0.1, f(x) lies in the range 14.21 to 15.81. Let's try delta = 0.05. This means that x lies between 3.95 and 4.05, for which $f(3.95) = 3.95^2 - 1 = 14.6025$ and $f(4.05) = 4.05^2 - 1 = 15.4025$. Since 14.6025 is greater than the desired 14.5, and since 15.4025 is also smaller than the desired 15.5, we were successful: delta = 0.05 works. We can achieve epsilon = 0.5 by choosing delta = 0.05. Note that this isn't the only possibility. We could choose any smaller value for delta, like 0.03 or like 0.0001, and that will work, too. We might even be able to choose 0.54 or 0.6 for delta (we'd have to calculate f(x) at each value to see), but it doesn't matter. In calculus, we don't need to find an optimal value for delta. We just need to know whether or not we can find one that works.

We've really done more arithmetic than necessary. In calculus, students don't usually do that much arithmetic. Generally, more of the work is done using symbols than numbers. We've worked out numerical examples to help illustrate the ideas and make them seem a bit more tangible.

Now we're ready to see how delta and epsilon are really used. If you are given any (nonzero) value for epsilon, is it always possible to find a (nonzero) value of delta that works? By "that works," we mean that for any value of x allowed by delta, the (absolute values) of the difference between f(x) and L is always less than epsilon. If this is always possible, the limit exists and it is equal to L.[44] If it isn't always possible (meaning that there is at least one exception), the limit does not exist.

We'll demonstrate this for a simple example. The example is simple in the sense that the function $g(x) = 3x + 8$ is pretty simple and straightforward. However, we'll

[44] When a student is given a function and asked to find the limit as x approaches a particular value, part of the challenge is also to determine the value of L.

do this the way it's often taught in calculus, using symbols. This might make it seem rather abstract. Indeed, many students who are taking a calculus course struggle with delta and epsilon notation, so if you find it difficult, that's to be expected. Since this is a conceptual book, one hope is that you can appreciate what calculus students go through, and hopefully you can glean a few of the main ideas that are going on. We'll finally use numbers at the end to try to make it seem a bit more tangible, and after we finish we'll summarize the main ideas for you. Hopefully, this will help.

Consider the function $g(x) = 3x + 8$. Let's find out if the limit exists as x approaches 2 and, if so, what the limit is. Which value should we try to use for L? Since $g(x)$ is finite when $x = 2$, it seems logical to try $g(2) = 3(2) + 8 = 6 + 8 = 14$. (And, in fact, this will work.) So we'll try letting L = 14.

The function is $g(x) = 3x + 8$, we want x to approach 2, and we're trying L = 14. The procedure says to start out by thinking about epsilon. What is epsilon? Recall that epsilon is a small deviation from L = 14. We're interested in the values of x where the difference between $3x + 8$ and L = 14 is smaller than epsilon. If x doesn't equal 2 exactly (and it won't, since x is approaching 2), then $3x + 8$ and L = 14 will be different. If we subtract 14 from $3x + 8$, we get $3x + 8 - 14 = 3x - 6$. This means that we want $|3x - 6|$ to be smaller than epsilon (where the vertical lines mean to take the absolute value of $3x - 6$, which means that if $3x - 6$ turns out to be negative, we'll discard the minus sign). Note that we're working with symbols here, rather than doing arithmetic. That's the way it is taught in calculus. We have an inequality for epsilon in terms of the independent variable: $|3x - 6| <$ epsilon. (The symbol < is the less than sign. It means that $|3x - 6|$ must be less than epsilon.)

Now that we have an expression for epsilon, our goal is to relate this to delta. Since we want x to approach 2, in this example delta is a small deviation from 2. If we're considering values of x for which $|3x - 6| <$ epsilon, this corresponds to values of x for which $|x - 2| <$ delta. These two inequalities are very similar. If we multiply both sides of delta's inequality by 3, we get $|3x - 6| < 3$ delta (where we used the distributive property of algebra to multiply the 3 with both the x and the 2). Compare $|3x - 6| <$ epsilon with $|3x - 6| < 3$ delta to see that we need epsilon to be 3 times delta or, equivalently, we need delta to be one-third of epsilon.

No matter which numerical value you use for epsilon, if delta is one-third the size of epsilon, this will automatically satisfy the criteria for the existence of the limit. That is, if delta is one-third of epsilon, this guarantees that the difference between $g(x)$ and L = 14 will be no greater than epsilon for any value of x from $(2 -$ delta$)$ to $(2 +$ delta$)$. If you really prefer to see it with numbers, let epsilon = 0.3. Then delta =

0.1 (that is, one-third of epsilon). We're choosing a value of x from 1.9 to 2.1 (since $2 - 0.1 = 1.9$ and $2 + 0.1 = 2.1$). The corresponding values of $g(x)$ are $g(1.9) = 3(1.9) + 8 = 5.7 + 8 = 13.7$ and $g(2.1) = 3(2.1) + 8 = 6.3 + 8 = 14.3$. All values of $g(x)$ in this example lie in the range 13.7 to 14.3 when x can vary from 1.9 to 2.1. Therefore, the limit exists and equals $L = 14$. If this seems like a lengthy solution, it's because we broke it down into several steps and added a lot of explanation. Calculus students solve this problem (using symbols only) with just a couple of lines of algebra.

Here's a recap of our solution. We started with $g(x) = 3x + 8$. We want to find the limit as x approaches 2. We used the delta and epsilon procedure to see if a limit exists with $L = 14$. We started out by considering how the value of x affects epsilon. The formula $g(x) = 3x + 8$ tells us how x affects g. Since we want epsilon to be smaller than the difference between $3x + 8$ and $L = 14$, we subtracted 14 to get $|3x - 6| <$ epsilon. A corresponding value of delta would satisfy $|x - 2| <$ delta (based on how delta is defined, since x is approaching 2). By comparison, we can always satisfy both inequalities if delta is one-third of epsilon, which shows that the limit exists and is equal to $L = 14$. This is the formal way to prove that $L = 14$ is the limit of $g(x)$ as x approaches 2. If you don't fully understand this, that's fine. The delta and epsilon procedure doesn't appeal to everyone. If you can understand how to find the limit using the informal method (using arithmetic) that we used in the previous chapter, that will be good enough for the purpose of this conceptual book.

$$y(x) = 0 \text{ if } x < 0$$
$$y(x) = 1 \text{ otherwise}$$

Let's try an example where the limit doesn't exist. Consider the step function above. We'd like to find the limit of $y(x)$ as x approaches zero, if it exists. We know from the previous chapter that the limit doesn't exist because the limit as x approaches zero from below is 0, whereas the limit as x approaches zero from above is 1; since the limits from below and above are different, the limit of $y(x)$ as x approaches zero does not exist. However, in this chapter we'll explore how to demonstrate that the limit does not exist using delta and epsilon.

Let's pretend that we don't know whether or not the limit as x approaches zero of $y(x)$ exists. If it does exist, we'll call this limit L. Let epsilon equal 0.2. (It turns out that in this example any value less than 0.5 will suffice.) Based on how epsilon is defined, this means that we need to consider values of x for which $|y(x) - L| <$ epsilon, meaning that $|y(x) - L| < 0.2$. If x is negative, then $y(x) = 0$ and we need $L < 0.2$. (Why? Zero minus L equals minus L, and the absolute value of minus L is just L) On the other hand, if x isn't negative, then $y(x) = 1$ and we need $|1 - L| < 0.2$. Both inequalities are needed. Since delta is a small deviation about zero (since x approaches zero in this problem), delta must allow for both positive and negative values of x.[45] This means we need both inequalities to hold. If the limit exists, we need L < 0.2 to be true and we need $|1 - L| < 0.2$ to be true. Is it possible to find a value of L such that both inequalities will be true? No. For example, suppose that L = 0.15. This value of L satisfies the first inequality, since 0.15 < 0.2. But 1 − L = 1 − 0.15 = 0.85 doesn't satisfy the second inequality, since 0.85 isn't less than 0.2. It can be shown[46] in general that if L < 0.2, then $|1 - L|$ won't be less than 0.2. Therefore, the limit of $y(x)$ as x approaches zero does not exist.

One might wonder how delta and epsilon can be used if x grows infinite. Delta is a small deviation from the value that x approaches. If we take a limit as x goes to infinity, how do you make a small deviation about infinity? Good question! The answer is that you can't. In this case, the language is modified.

If we wish to find the limit of $f(x)$ as x goes to infinity, as before we begin by considering a value of epsilon, which is a small deviation from L, if it exists. This means that we want values of x for which the difference between $f(x)$ and L is smaller than epsilon. For a given value of epsilon, is it always possible to find a large enough value of x so that for all values of x larger than that, $f(x)$ and L will satisfy this requirement?[47] (This is similar to the case where x approaches a finite value, except that now we're not working with delta. In the case where x goes to infinity, delta doesn't make any sense. You can't make a small deviation from infinity.)

For example, consider the function $f(x) = 2/x$ in the limit that x goes to infinity. As x gets larger, $2/x$ gets smaller, so it seems reasonable to try L = 0. That is, we expect $f(x)$ to approach zero as x goes to infinity. We'll apply the above definition to see if

---

[45] We need delta < $|x - 0|$, which simplifies to delta < $|x|$. It may help to compare with the previous example, where we had $|x - 2|$. In this example, x approaches zero (rather than 2).

[46] This footnote is intended for students who are really fluent with algebra. Students who are well-versed in the algebra of inequalities and absolute values may know that $|1 - L| < 0.2$ can be expressed as −0.2 < 1 − L < 0.2. Now add L to each side of each inequality to get L − 0.2 < 1 < L + 0.2. Add 0.2 to both sides of L − 0.2 < 1 to get L < 1.2, and subtract 0.2 from both sides of 1 < L + 0.2 to get 0.8 < L. Combining L < 1.2 with 0.8 < L, we get 0.8 < L < 1.2. No value of L can possibly satisfy L < 0.2 and also satisfy 0.8 < L < 1.2. For example, if L is less than 0.2, it definitely won't be greater than 0.8.

[47] If you want to consider a limit where x approaches negative infinity, we need to revise this language somewhat. To keep things simple, in this chapter, we'll only consider x going to positive infinity (not negative infinity). The two cases are very similar.

indeed L does exist and equals zero. Recall that epsilon is a small deviation from L. Similar to our previous example, we are interested in values of x for which the difference (in absolute values) between $f(x) = 2/x$ and L is less than epsilon. Since $f(x) - L = 2/x - 0 = 2/x$, this means that we want $|2/x| <$ epsilon. Since x is going to infinity, clearly x will be positive in the region of interest, so we will drop the absolute values: $2/x <$ epsilon. Multiply by x on both sides to get $2 < x$ times epsilon, and divide by epsilon on both sides to get 2/epsilon $< x$. This shows that for all x greater than 2/epsilon, $f(x) = 2/x$ and L = 0 will satisfy our requirement for epsilon. We met the criteria of the previous paragraph, which shows that this limit exists and equals zero. For example, if we want epsilon to be 0.1, then any x greater than 2/0.1 = 20 will make the difference between $f(x) = 2/x$ and L = 0 smaller than 0.1. For example, let's try $x = 25$. When $x = 25$, we get f(25) = 2/25 = 0.08, which is less than 0.1. The main idea is that no matter which (positive) numerical value you choose for epsilon, any x greater than 2/epsilon will guarantee that $|f(x) - L|$ is smaller than epsilon, which proves that the limit exists and equals L = 0.

Don't confuse the limit of f(x) = $2/x$ where x goes to infinity with the limit of f(x) = $2/x$ where x goes to zero (or some other value). The value that x approaches is very important. In the previous paragraph, we found that the limit of f(x) as x goes to infinity is zero. In contrast, the limit of f(x) as x approaches zero does not exist. (This is similar to an example from Chapter 6. When x approaches zero from above, $2/x$ approaches positive infinity, but when x approaches zero from below, $2/x$ approaches negative infinity, and since the limits from below and above are different, the limit does not exist as x approaches zero.) Yet another limit regarding f(x) = 2/x is that the limit of f(x) is 2 as x approaches one.

By now, you should have some idea of what a limit means, and some understanding of how we evaluated limits in the previous chapter (by trying different values of x and looking for a pattern). If so, you're ready to move onto Chapter 8 (even if you have some uncertainty about Chapter 7).

Ready for a killer pop quiz? (Ch. 7)

If you finished reading Chapter 7, you get an A. (That wasn't too bad, was it?)

8 What is a derivative?

This is arguably the most important chapter of this book. Two of the most practical skills that calculus teaches are how to find derivatives and integrals (and one generally needs to learn how to find derivatives before one is taught how to find integrals). Learning about derivatives is about as important to calculus as learning how to isolate an unknown is important to algebra. The good news is that derivatives are fairly straightforward and can be relatively easy to understand.

We will develop the concept of a derivative using a visual approach. This way, you will be able to 'see' what a derivative means. Before we begin, we will quickly review two important concepts: slope and tangent lines.

Slope is a measure of how steep a line is.[48] For example, consider the straight lines illustrated below. The horizontal line at the left has zero slope; it isn't steep at all. The second line from the left has a little slope; it is only slightly steep. The line in the center has more slope. The second line from the right has even more slope; it is fairly steep. The vertical line at the right has infinite slope; it's as steep as it can get.

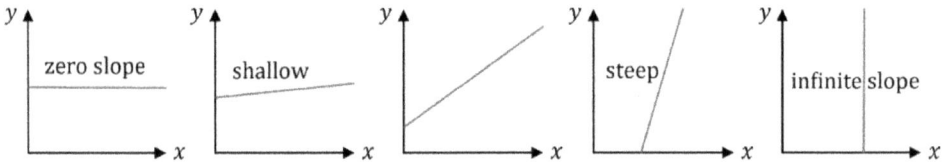

The figure above and the previous paragraph provided a qualitative description of slope. In mathematics, we define **slope** quantitatively as the rise divided by the run. For example, consider the straight line below. A right triangle has been drawn using horizontal and vertical legs. The **rise** is the height of the vertical side, and the **run** is the length of the horizontal side.

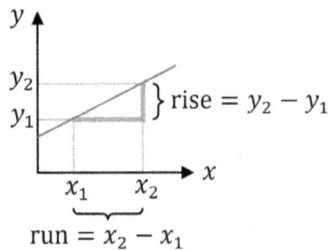

---

[48] As we'll discover, you can also talk about the slope of a curve at various points. In that case, slope is a measure of how steep the tangent line is. But since we haven't yet explained what a tangent line is, we'll begin by discussing the slope of a straight line.

The bottom left corner of the triangle has coordinates (x_1, y_1) and the top right corner of the triangle has coordinates (x_2, y_2). The (x, y) coordinates of any point tell you where the point is relative to the origin (which is where the x-axis and y-axis cross). The value of x tells you how far the point is to the right of the origin, while the value of y tells you how far the point is above the origin. The rise is equal to $y_2 - y_1$, and the run is equal to $x_2 - x_1$. Since the slope of the line equals the rise over the run, the formula for the slope is $(y_2 - y_1)/(x_2 - x_1)$. These details are typically taught in an algebra course. Before you worry that you might need to do calculations with numbers using these formulas, for our purposes, the goal is to understand what this formula means in terms of words and ideas. Primarily, you want to remember that slope is a measure of how steep a line is, and that slope equals rise over run. It will be handy (especially, in Chapter 14) to know that a point with coordinates (x, y) is located x units to the right of the origin and y units above the origin, and to understand that the formula below represents the rise (which is vertical) over the run (which is horizontal), as illustrated above.

$$\text{slope} = (y_2 - y_1)/(x_2 - x_1)$$

In this equation, the numerator $(y_2 - y_1)$ is the rise, while the denominator $(x_2 - x_1)$ is the run.

$$\text{slope} = \frac{\text{rise}}{\text{run}} = \frac{y_2 - y_1}{x_2 - x_1}$$

The bigger the value of the slope, the steeper the line is. For example, the line on the left below has a slope of 0.5, the line in the center has a slope of 1, and the line on the right has a slope of 2. For the center line, which has a slope of 1, the rise equals the run. Every time the line in the center goes one square to the right, it also goes up one square. In contrast, for the line on the right, which has a slope of 2, every time the line goes 1 square to the right, it goes up 2 squares. For the line on the right, the rise is 2 when the run is 1, so the slope $= 2/1 = 2$. It will be helpful to know that a steeper line has a greater slope, and that a horizontal line has zero slope.

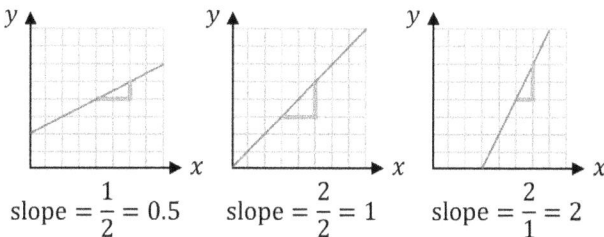

slope $= \dfrac{1}{2} = 0.5$ slope $= \dfrac{2}{2} = 1$ slope $= \dfrac{2}{1} = 2$

A straight line has constant slope. For example, consider the straight line drawn below. Either right triangle gives the same value of the slope. In fact, any right triangle with horizontal and vertical sides (where the hypotenuse is a segment of the line) will yield the same slope. That's because the slope of the line is constant. In the smaller right triangle below, the rise is 1 and the run is 2, such that the slope is 1/2. In the larger right triangle below, the rise is 2 and the run is 4, for which the slope is 2/4 = 1/2. It's the same, since 2 divided by 4 equals 0.5 (the same as 1 divided by 2).

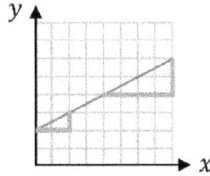

The general equation for a straight line may be expressed in the form y = mx + b, where m represents the slope and b is the y-intercept. (The y-intercept, b, is the value of y when the line crosses the vertical axis.)

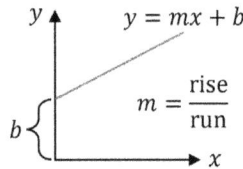

One more point about the slope of a straight line that is worth noting is that it can be negative. If the line rises upward as it goes to the right, the slope is positive, like the line below on the left. If instead the line falls downward as it goes to the right, the slope is negative, like the line below on the right. The slope of a horizontal line, on the other hand, is neither positive nor negative; a horizontal line has zero slope.

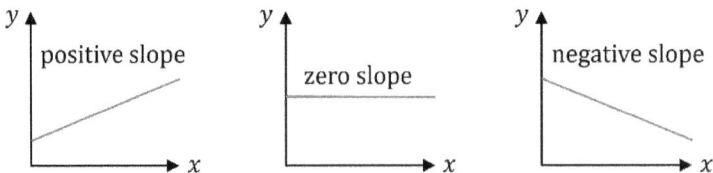

In calculus, we're concerned with the slope of a curve, rather than a straight line. Along a curve, the slope changes. (Something about this statement should sound familiar. Earlier in this book, we described calculus as the mathematics of change.) When we refer to the slope along a curve, we mean the slope of a line that is tangent to the curve. So we'll take a moment to discuss what a tangent line is.

When teaching physics or calculus, if the instructor asks students what a tangent line is, it's surprisingly common for the answer to be incorrect (or at least imprecise). Most students who take physics or calculus have a fairly good understanding of what a tangent line looks like. If you draw a curve, mark a point on the curve, hand a ruler to a student, and ask the student to draw a tangent line at that point, most students would do a fairly good job without needing to review the definition of tangent. Yet if you ask them to explain what a tangent line is in words, it's very common for the answer to be incorrect or imprecise. And there's a reason for it.

The most common answer, just going by memory of what they learned in previous math courses, is this: "The tangent line touches the curve at a single point." This explanation was probably correct in the context in which the student learned about tangent lines, but this definition isn't correct in general, as we will see now. We'll improve upon the definition above.

Let's examine what's wrong with the above quotation. There are two problems with that definition, which are illustrated below. In the curve on the left, a line has been drawn which touches the curve at a single point, but this line is most definitely NOT tangent to the curve. Almost every student would tell you that the line on the left curve below is NOT a tangent line, even the students who instinctively believe that a tangent line is defined as a line that touches the curve at a single point. That's one problem with the quotation above. In the curve on the right, a line has been drawn which is tangent to the curve at one point, but actually touches the curve at a second point. This shows that it's possible for a tangent line to intersect a curve at multiple points. That's the second problem with the quotation above.

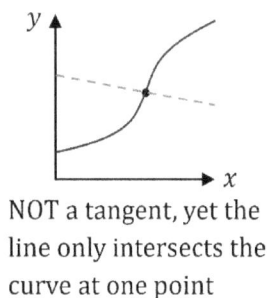

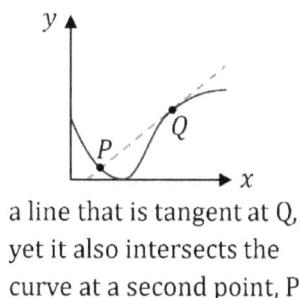

NOT a tangent, yet the line only intersects the curve at one point

a line that is tangent at Q, yet it also intersects the curve at a second point, P

Why do so many students remember the definition of a tangent line as a line that intersects the curve at a single point? The answer is easy. Students learned that defini-tion in some math class where the teacher was discussing circles. It could have been in algebra, geometry, or trigonometry. Surely, the instructor had a circle on the board, and correctly explained that, for a circle, a tangent line touches the circle at a single point. As shown below, a line that touches a circle at a single point is tangent to the circle. [49] The problem with the definition of touching a curve at a single point is that it doesn't apply to all curves, like cubics. The pair of figures above demonstrate why this common definition doesn't suffice. (And yet this same imprecise definition of a tangent line is actually given in some very popular dictionaries.) So now we'll adapt the definition of a tangent line so that it applies more generally. [50]

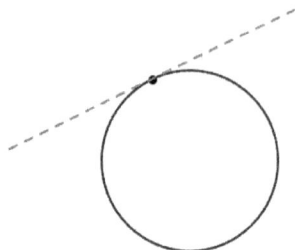

In general (for smooth sections of curves without discontinuities), a line is **tangent** to a curve at a particular point on the curve if the line matches the slope of the curve at that point and if the line touches the curve at a single point in the desired section of arc. Two important points have been added to the common but imprecise definition. First, the tangent line must have the same slope as the curve at the point where it is tangent to the curve. This wording disallows the left diagram below. Second, we added that the tangent line touches the curve only once in the desired section of arc. What do we mean by a 'desired section of arc'? Imagine an ant crawling forward along the curve. If the ant is turning to the left presently, that's one section of arc. If later the ant is turning to the right, that's a new section of arc. Later yet the ant might be turning to the left again, which would be a third section of arc. This language allows the tangent line to intersect the curve once locally, and then intersect it again where the curve has changed direction, as in the right diagram below.

---

[49] A line that intersects a circle at two points is called a secant (which isn't tangent to the circle). A secant is an extension of a chord, which is a line segment joining two points that lie on the circumference.

[50] Even this definition may have some limitations. If the curve isn't smooth or if the curve has discontinuities, there may be special points where the concept of a tangent seems to be problematic. This is important in calculus. The derivatives that we're about to learn about apply to sections of smooth curves without discontinuities.

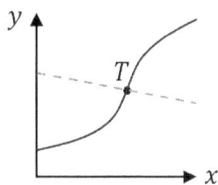

NOT tangent because
the slope doesn't match
the curve's slope at T

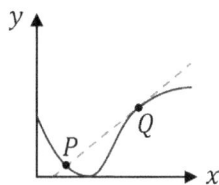

tangent at Q, but not P;
points P and Q are not on
the same section of arc

The specific language needed to explain precisely what a tangent line is helps to make the definition technically correct, but what really matters is if you understand how to draw a tangent line. Most students can actually do the part that really matters, so it's not so troublesome if their intuitive definition of a tangent allows for the loopholes shown above.

How do you draw a tangent line? Imagine that a smooth curve has been drawn on a chalkboard. One particular point, labeled P, is indicated on the curve. Such a curve is illustrated below. You wish to draw a line that is tangent to the curve at that point. You have a meterstick and a piece of chalk (or you can use the curve shown below, a ruler, and a pencil). Consider how you would draw a line that is tangent to the curve at point P.

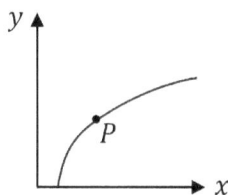

Hold the meterstick (or ruler) up the curve so that point P lies along the edge of the meterstick (obviously, the edge you intend to draw on). Now rotate the angle of the meterstick (or ruler) so that P still lies on the edge of the meterstick and so that the angle of the meterstick matches the steepness of the curve at point P. See the middle diagram below. (The left diagram below is incorrect because the line is less steep than the curve at P, while the right diagram below is incorrect because the line is steeper than the curve at P. The right case also intersects the curve at two points locally, unlike the other cases.)

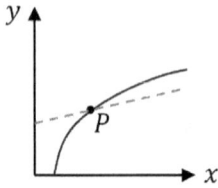

NOT tangent because the line is less steep than the curve at P

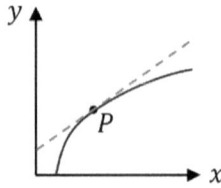

tangent at P

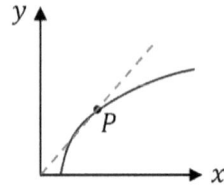

NOT tangent because the line is steeper than the curve at P

In Chapter 5, we learned about the tangent function, which relates two sides of a right triangle. In this chapter, we learned about a line that is tangent to a curve. You might wonder why we use the same term 'tangent' in both cases. You can understand this by considering the tangent line drawn below. We formed a right triangle by adding horizontal and vertical lines. This right triangle helps us find the slope of the tangent line.

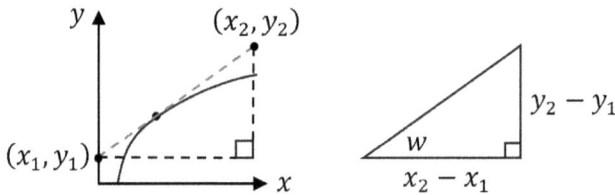

Our formula for the slope of a straight line is

$$\text{slope} = (y_2 - y_1)/(x_2 - x_1)$$

Since the tangent line is a straight line, this is the slope of the tangent line. The rise, $y_2 - y_1$, is opposite to angle w in the diagram above, while the run, $x_2 - x_1$, is adjacent to angle w. From our definition of the tangent function (Chapter 5), the tangent of w equals the opposite side over the adjacent side, which gives us

$$\text{tangent } w = (y_2 - y_1)/(x_2 - x_1)$$

Compare the two equations above to see that the tangent of angle w is equal to the slope of the tangent line. Thus, it should be no surprise that the same term 'tangent' is used in both cases.

$$\text{slope} = \frac{y_2 - y_1}{x_2 - x_1} \quad , \quad \tan w = \frac{y_2 - y_1}{x_2 - x_1} \quad , \quad \tan w = \text{slope}$$

We've reviewed what we mean by a tangent line, and how to find the slope of a straight line. Now we will apply these concepts to learn a little calculus.

Imagine that we have a function $y(x)$, which is a smooth curve without discontinuities, and we wish to find the slope of the tangent line for a given value of x. Calculus helps us figure this out without having to make a graph and draw a tangent line. This calculation involves what is known as a derivative. Since this is a conceptual book, we won't do the calculation purely with algebra. We'll use the graph as a visual aid to develop what a derivative means and how to find it (at least, for relatively simple cases).

To learn what a derivative is, consider the following problem. We are given the formula for some function $y(x)$. We're interested in the section of the curve beginning at the point (x_i, y_i) and ending at the point (x_f, y_f). The subscript i stands for initial and the subscript f stands for final. The function is smooth and continuous between these two points. At some point P along the curve between these two points, we would like a formula that gives us the slope of the tangent line. For example, we might be given $y(x) = 3x^2$ and wish to know the slope of the tangent line when $x = 4$, we might be given $y(x) =$ square root of x and wish to know the slope of the tangent line when $x = 9$, or we might be given $y(x) = \cos(x)$ and wish to know the slope of the tangent line when $x =$ pi over 6 (that is, when $x = \frac{\pi}{6}$).

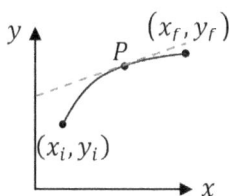

Goal: Find the slope of the line that is tangent to the curve at P.

If the function were a straight line, this would be easy. We would just find the slope of the straight line connecting (x_i, y_i) to (x_f, y_f). This slope would be

$$\text{slope} = (y_f - y_i)/(x_f - x_i)$$

However, if the function is a curve, the actual slope at the desired point P may be considerably different from the slope of the line that joins (x_i, y_i) to (x_f, y_f). The illustrations below show three of the many possible functions that may join (x_i, y_i) to (x_f, y_f), and in each case the slope of the tangent line at P is quite different from the slope of the line that joins (x_i, y_i) to (x_f, y_f). The line that joins (x_i, y_i) to (x_f, y_f) provides a measure of the average slope, but it's NOT in general a very good measure of the slope of the tangent line at P.

One line is tangent to point P, whereas the other line joins (x_i, y_i) to (x_f, y_f). In general, the slopes of these two lines can be considerably different.

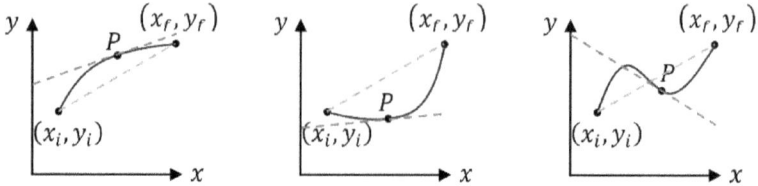

What if we draw a straight line that joins two points that are closer to P? The left diagram below chooses two points closer to P. In the middle diagram, the points are even closer to P. In the right diagram, they are closer yet. The closer these two points are to P, the less room there is for the curve to change considerably in between them, suggesting that the slope of the line that joins those points is more likely to be close the slope of the tangent line at P.

The closer the two points are to point P, the better the slopes of the two lines match; the line through P is the tangent line.

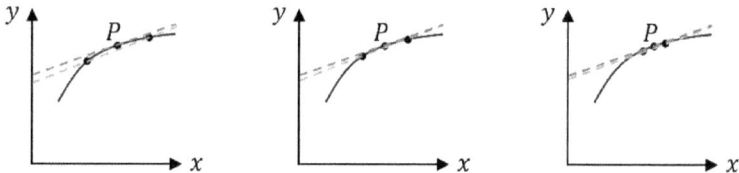

This is the very spirit with which the concept of a derivative is developed. (We keep mentioning this word 'derivative,' but haven't yet explained what it is. Be patient, we'll get there. You might have guessed that it has something to do with the slope of the tangent line, and, if so, you'd be correct.) The closer the two points on the curve are to one another (with point P in between them), the closer the line that joins these two points will be to the actual tangent line at P.

This is a limit (Chapters 6-7). If the coordinates of the two points are (x_1, y_1) and (x_2, y_2), then the slope of the line joining these two points is

$$\text{slope} = (y_2 - y_1)/(x_2 - x_1)$$

In the limit that the run, $x_2 - x_1$, approaches zero, the line joining (x_1, y_1) and (x_2, y_2) will be the line that is tangent to P, and the slope given by the equation above will be equal to the slope of the tangent line. We don't want the run $x_2 - x_1$ to be exactly equal to zero because then we would have a division by zero problem, but we want $x_2 - x_1$ to be as small as possible while being nonzero.

That is, we want $x_2 - x_1$ to be **infinitesimal** (that is, as small as possible without being exactly zero). If the quantity $x_2 - x_1$ is infinitesimal, it will be a differential element. A **differential element** is an infinitesimal amount of a quantity; in this case, it's an infinitesimal amount of the horizontal variable x. If the run is infinitesimal, the rise will also be infinitesimal. We use the notation dx to represent a differential element along the horizontal coordinate x. That is, dx is infinitesimal. Similarly, dy is a differential element along the vertical coordinate y, meaning that dy is infinitesimal.

When we take the limit of the slope of the tangent line as $x_2 - x_1$ shrinks to zero, we call this quantity a **derivative** of y with respect to x. The notation for this is dy/dx. (The reason behind this notation has to do with the differential elements mentioned in the previous paragraph.) Specifically, dy/dx represents a derivative of y with respect to x. When we evaluate the derivative at point P, the value of the **derivative represents the slope of the tangent line** at P. The derivative dy/dx is **defined as the limit** of $y_2 - y_1$ as the quantity $x_2 - x_1$ shrinks to zero. Recall that point P lies between points (x_1, y_1) and (x_2, y_2), and the line that joins (x_1, y_1) and (x_2, y_2) becomes the tangent line in the limit that $x_2 - x_1$ approaches zero. Recall our original goal. We'd like to find a formula for the derivative, which we will call dy/dx, given the formula for the function $y(x)$.

To find this formula, we'll begin by combining ideas that we already know at this point. The derivative of y with respect to x, denoted as dy/dx, is the limit of $y_2 - y_1$ divided by $x_2 - x_1$ as the quantity $x_2 - x_1$ approaches zero. (You should recognize that $y_2 - y_1$ divided by $x_2 - x_1$ is the slope of the line that joins the two points.) Next, we'll rewrite this same statement with revised notation. Why introduce new notation? The original notation, with x_1, y_1, x_2, and y_2 makes sense in the original context of joining the points (x_1, y_1) and (x_2, y_2). But now that we're taking a limit as something approaches zero, it would seem simpler to have a single symbol approach zero instead of taking a limit as the quantity $x_2 - x_1$ approaches zero. So that's why. (For this very reason, this is a standard change in notation common in many calculus texts that introduce the derivative by considering limits.)

First, note that $y(x)$ is a function of x, such that y_1 really means $y(x_1)$ and y_2 really means $y(x_2)$. That is, whether we write y_1 or write $y(x_1)$, either way this means to evaluate the function $y(x)$ when x is equal to x_1, and similarly for y_2 or $y(x_2)$. So instead of writing $y_2 - y_1$ in the numerator of the formula, we could just as well write $y(x_2) - y(x_1)$. This is one change that we will make.

A second change that we will make is that we will just write h instead of $x_2 - x_1$. When you see h, it's equal to $x_2 - x_1$. It will just be simpler to take the limit as h approaches zero than to consider a limit where $x_2 - x_1$ approaches zero. Since $x_2 - x_1$ = h, we may add x_1 to both sides and rewrite this as $x_2 = x_1 + h$. This means that y(x_2) is equivalent to y(x_1+h).

With these changes, we're ready to rewrite our original definition of the derivative in terms of a limit. Originally, we said that the derivative of y with respect to x is the limit of $y_2 - y_1$ divided by $x_2 - x_1$ as the quantity $x_2 - x_1$ approaches zero. Now we may rewrite this as follows. The derivative of y with respect to x is the limit of y(x_1+h) – y(x_1) divided by h as the quantity h approaches zero.

In fact, we'll go a step further. Observe the current statement only involves x_1 (and not x_2 explicitly). Since there aren't two different x's to keep track of anymore, we might as well drop the subscript and simply write x instead of x_1. So now we will state the derivative as the following.

The derivative of y with respect to x is the limit of y(x +h) – y(x) divided by h as the quantity h approaches zero. This derivative is represented by $\frac{dy}{dx}$.

That's the equation that calculus students use to find a formula for the derivative $\frac{dy}{dx}$ when they are given a formula for y(x). And what does the derivative mean? If you make a graph of y as a function of x, the value of the derivative tells you the slope of the tangent line at a particular value of x.

We'll now go through some examples to see how you can use the definition of the derivative above to find the slope of a tangent line from a formula for y(x). Beware that the first example will work out a little calculus, using some algebra, but the first example is important as it will also show you how the answer for the derivative is the slope of the tangent line. Seeing a little calculus will help you see and appreciate a bit of what calculus students actually do, but if you don't care for the algebra involved or find it difficult to follow, then focus on the other side of the example, striving to see how the derivative in each case relates to slope. It's the second part, trying to under-stand what a derivative is in terms of words and pictures, that's the main goal of this book. And if you don't care for the algebra that you see in the first example, there's good news. The remaining examples scarcely involve any algebra. If that's what you prefer, you have something to look forward to once you get through the first example (which provides a taste of real calculus).

As our first example, consider the function y(x) = x^2. A graph of y as a function of x for the curve y = x^2 is called a parabola; this graph is shown below. To use the definition above, first we find y(x+h), which means to replace x with x + h in the formula y(x) = x^2. When we do this, we get y(x+h) = $(x+h)^2$. Those who are fluent in algebra know that $(x+h)^2$ = x^2 + 2xh + h^2. The formula for the derivative says to subtract y(x) from y(x+h), and then divide by h. When we subtract y(x+h) – y(x), the x^2 terms cancel out, and we're left with 2xh + h^2. Then when we divide by h, we get 2x + h. Finally, the definition for the derivative says to take the limit as h approaches zero of 2x + h. When h goes to zero, 2x + h simply becomes 2x. Our final answer is that a derivative of y = x^2 with respect to x equals 2x. We write the answer as $\frac{dy}{dx}$ = 2x, where $\frac{dy}{dx}$ is our general notation for a derivative of y with respect to x, and where 2x turned out to be the answer in this example.

$$\frac{dy}{dx} = \lim_{h \to 0} \frac{y(x+h) - y(x)}{h} = \lim_{h \to 0} \frac{(x+h)^2 - x^2}{h} = \lim_{h \to 0} \frac{x^2 + 2xh + h^2 - x^2}{h}$$

$$= \lim_{h \to 0} \frac{2xh + h^2}{h} = \lim_{h \to 0} \frac{(2x+h)h}{h} = \lim_{h \to 0}(2x + h) = 2x$$

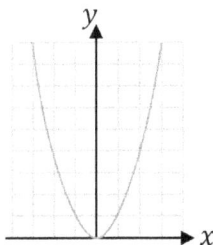

(If this graph scares you, Footnote 51 gives you a quick refresher of how coordinates relate to graphs.[51] Footnote 52 discusses the parabola in the graph above.[52])

---

[51] Each point along a curve has (x, y) coordinates. The origin is the point where the x- and y-axes cross; it is the point (0, 0). Given a point (x, y), the first coordinate (x) tells you how far it is to the right of the origin, while the second coordinate (y) tells you how far it is above the origin. For example, for the graph of the parabola shown, the top right point that you can see on the curve is the point (3, 9); this point is 3 squares to the right of the origin and 9 squares above it. (Try counting the squares.) Of course, this isn't the highest point on the curve; the curve continues up and to the right infinitely. The graph shown is only 9 squares high, so any point where y would be greater than 9 doesn't show on this graph.

[52] Use the equation y = x^2 to understand this graph. When x = 0, we get y = 0^2 = 0, so one point on the graph is the origin, (0, 0). When x = 1, we get y = 1^2 – 1, so another point is (1, 1). When x = 2, we get y = 2^2 = 4, so another point is (2, 4). When x = 3, we get y = 3^2 = 9, so another point is (3, 9). See if you can identify each of these points on the graph shown. (The previous footnote should be helpful.) Other points include (–1, 1), (–2, 4), and (–3, 9); when

For those who got lost in the algebra, here is the basic idea of what we did. First, we evaluated the function $y(x) = x^2$ at a point that is slightly larger than x; that point is x + h. (This step involved some algebra, since we had to multiply x + h by itself.) Then we subtracted y = x^2 and divided by h, because that's what our definition of a derivative in terms of a limit says to do; it says to find $y(x+h) - y(x)$ and divide by h. When we did this, we arrived at the expression $2x$ + h. Finally, when we took the limit of $2x$ + h as h approaches zero, the h disappeared in the limit and all that remained was $2x$. This process tells us that the derivative of y = x^2 with respect to x is $2x$. We write the answer for this derivative using the notation $\frac{dy}{dx} = 2x$.

Let's see how the answer, that the derivative is $2x$, relates to the slope of the tangent lines of the given function, y = x^2. Examine the graph of the parabola y = x^2 shown above. Think about drawing tangent lines at various points on the curve. To help you visualize this, we've added a handful of tangent lines in the graph below.

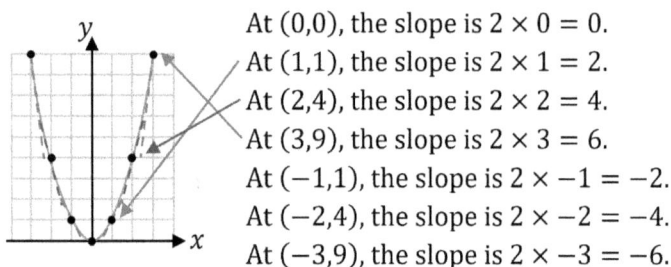

At (0,0), the slope is $2 \times 0 = 0$.
At (1,1), the slope is $2 \times 1 = 2$.
At (2,4), the slope is $2 \times 2 = 4$.
At (3,9), the slope is $2 \times 3 = 6$.
At (−1,1), the slope is $2 \times -1 = -2$.
At (−2,4), the slope is $2 \times -2 = -4$.
At (−3,9), the slope is $2 \times -3 = -6$.

The bullet points below use the formula $dy/dx = 2x$ to evaluate the derivative and compare this derivative to the slope of the corresponding tangent line. These bullet points help you see that the derivative relates to the slopes of tangent lines.

• When x = 0, the derivative is $dy/dx = 2x = 2(0) = 0$. At x = 0, the slope of the tangent line is zero, which means that the tangent line is horizontal at this point. This is the point at the bottom of the parabola in the figure above. It's a special point. As we'll explore in Chapter 10, when a curve makes a relative minimum or maximum, the derivative is zero at that point; in this example, this is the minimum value of y for the parabola.

• When x = 1, the derivative is $dy/dx = 2x = 2(1) = 2$. At x = 1, the slope of the tangent line is 2. It's a positive slope where the rise is twice the run.

• When x = 2, the derivative is $dy/dx = 2x = 2(2) = 4$. At x = 2, the slope of the tangent line is 4. It's twice as steep compared to the tangent line at x = 1.

---

x is negative we get the same value for y as when x is positive. Why? Because x is squared: y = x^2. A negative value squared is positive, so $(-2)^2 = 4$ gives the same value for y as $2^2 = 4$, for example.

- When $x = 3$, the derivative is $dy/dx = 2x = 2(3) = 6$. At $x = 3$, the slope of the tangent line is 6. It's three times as steep compared to the tangent line at $x = 1$. If you examine the tangent lines when $x = 1, 2$, and 3 in the graph above, you should observe that the tangent lines are getting steeper as x increases, which agrees with the derivatives at $x = 1, 2$, and 3.
- When $x = -1$, the derivative is $dy/dx = 2x = 2(-1) = -2$. At $x = -1$, the slope of the tangent line is -2. This slope is negative and the rise is twice the run. Find this point on the graph above (it's left of the bottom). This tangent line runs down to the right because its slope is negative.
- When $x = -2$, the derivative is $dy/dx = 2x = 2(-2) = -4$. At $x = -2$, the slope of the tangent line is -4. This slope is negative and steeper than the tangent line at $x = -1$. If you examine the tangent lines when $x = -1, -2$, and -3 in the graph above, you should observe that the tangent lines are getting steeper as x becomes more negative.

If you ask a student who has taken calculus to find the derivative of $y = x^2$, most students who know how to do it probably **won't** solve the problem the way that we just did using the definition of a derivative as a limit. Students only use the definition of a derivative as a limit when they first learn about derivatives. After that, for the rest of the course (and any subsequent calculus courses that they may take, or other courses like physics where they apply calculus), students just use a quick-and-easy formula to find the derivative. The definition of a derivative as a limit is important to understand because it shows what a derivative really is and it helps to find the derivative when you come across a new kind of function. But the formulas for derivatives that don't involve limits are practical, making it easy to find derivatives. The practical formulas come from the definition of a limit.

For example, one practical formula that calculus students use for finding derivatives applies to polynomials like $y(x) = 6x^5$, $f(x) = x^2 - 2x + 3$, or $g(x) = x^8 - 5x^6 + 9x^4 - 2x^2 + 11$. This formula says that if $y(x)$ has the form ax^b, where the coefficient (a) and exponent (b) are both constants, then the derivative is $\frac{dy}{dx} = abx^{b-1}$. This formula says to multiply the coefficient by the exponent (that's the ab part) and reduce the exponent by one (that's the $b - 1$ part).

Let's verify that this formula works for our previous example, $y(x) = x^2$. Compare x^2 with the general form ax^b. What are a and b? The 'trick' here is that a = 1. Recall from algebra that $1x = x$. If you don't see a coefficient in algebra, it's implied that the coefficient is one.[53] The easy part when comparing x^2 with ax^b is to see that b = 2. Once you see that a = 1 and b = 2, just use the formula $dy/dx = abx^{b-1}$. This gives us $(1)(2)x^{2-1}$, which simplifies to $2x^1$. Another thing you need to recall from algebra is that $x^1 = x$. That is, raising a number to the first power doesn't change the number. So, our final answer is that $dy/dx = 2x$. We took a paragraph to explain every little detail, but calculus students solve this problem in one quick step.

As a second example, consider the function $f(x) = 2x^5 - 4x^3$. We can apply the same formula to each term. (Recall that the terms of an equation are separated by + signs, − signs, and = signs, so that $2x^5$ and $4x^3$ are two different terms on the right-hand side.) Compare $2x^5$ with ax^b to see that a = 2 and b = 5, which gives $abx^{b-1} = (2)(5)x^{5-1} = 10x^4$, and compare $4x^3$ with ax^b to see that a = 4 and b = 3, which gives $abx^{b-1} = (4)(3)x^{3-1} = 12x^2$. Since $f(x)$ has two terms being subtracted, df/dx will subtract these two answers. Our final answer for the derivative of $f(x)$ with respect to x is $\frac{df}{dx} = 10x^4 - 12x^2$. You may have noticed that we used $f(x)$ and df/dx in this example rather than $y(x)$ and dy/dx. The same symbol isn't always used for the function (or even the variable) in calculus.[54] Whatever letter is used for the function, we just go with it.

What if $y(x)$ is a straight line, like $y(x) = 3x - 2$? (How is this a straight line? The general formula for a straight line is $y = mx + b$. In this problem, we have $y = 3x - 2$. By comparison, $y = 3x - 2$ is a straight line with a slope of m = 3 and a y-intercept of −2.) Similar to the previous example, we will separate this into two terms. The first term is $3x$. Compare $3x$ with ax^b. It's easy to see that a = 3, but what is b? The 'trick' is that b = 1 because of the rule from algebra that $x^1 = x$. (When you don't see an exponent in algebra, it's implied that the exponent equals one.) Use a = 3 and b = 1

---

[53] Why? Because multiplying by one doesn't change a number. For example, if you have the number 4 and you multiply it by one, you get $(1)(4) = 4$. The 4 remains unchanged. This is true if we multiply any number by one. Since the variable x can be any number, if we multiply x by 1, it remains unchanged: $1x = x$. So in reverse, if we just see x all by itself, we could write $x = 1x$. That's why a coefficient of 1 is always implied in algebra when you don't see a co-efficient. (If you forgot what the term coefficient means, this word refers to the constant that multiplies a variable.)

[54] Often, there is no choice. If the same problem involves multiple functions, you can't give every function the same name. One function might be $y(x)$ = square root of x and another function might be $g(x) = \sin x$. If we called them both $y(x)$, it would be confusing. In the same problem, each different function needs its own unique name. So, it's good to be flexible and adapt to using different symbols.

to get $abx^{b-1} = (3)(1)x^{1-1} = 3x^0$. Now what is x^0? Recall from algebra that $x^0 = 1$ (as explained in Footnote 32) if x is nonzero. Using $x^0 = 1$, the derivative of $3x$ is simply $3(1) = 3$. Now we need the derivative of the second term, which is 2.[55] Compare 2 with ax^b. Wait a minute, where's the x? We'll use the rule $x^0 = 1$ again to introduce the x. That is, $2 = 2(1) = 2x^0$. Now compare $2x^0$ with ax^b to see that a = 2 and b = 0. This gives $abx^{b-1} = (2)(0)x^{0-1} = 0x^{-1} = 0$. (Zero times anything is zero, so when the exponent zero comes out to multiply the coefficient to form the derivative, we find that the derivative of the second term is zero.) Putting the two terms together, the derivative of y = $3x - 2$ is equal to $\frac{dy}{dx} = 3 - 0 = 3$.

We just did a lot of work to get something that we should have already known to begin with. What does a derivative do? It tells you the slope of the tangent line. The function y = $3x - 2$ is the equation for a straight line, not a curve. The slope of a straight line doesn't change. For any value of x on the straight line y = $3x - 2$, the slope is m = 3. When we take the derivative of a straight line, we just get the slope of the line, which is constant.

We actually learned two important rules for derivatives in the previous example. One thing we learned is that the derivative of $3x$ with respect to x equals 3. This is the part where the derivative tells us that any straight line has constant slope, meaning that dy/dx equals a constant if y is a straight line. A second thing that we learned in the previous example is that a derivative of 2 with respect to x equals zero. That is, **the derivative of a constant is zero**. Consider the equation y = 2. This is a (horizontal) straight line with zero slope and a y-intercept of 2. A line with zero slope is horizontal. For the special line y = 2, for any value of x the value of y is 2. Since the slope of a horizontal line is zero, the derivative of a constant equals zero.

Let's look at an example which at first glance might seem quite different. Consider the function f(x) = square root of $x = \sqrt{x}$. What is a derivative of f with respect to x? Are you thinking, "We haven't yet learned how to find the derivative of a square root?" If so, even if you don't know any more calculus than you've already read in this book, you'd be wrong. You actually do know how to find the derivative of a square root. You just need to know an important rule from algebra (and even if you've forgotten that rule from algebra, the rule that you need to know was stated in Chapter 5, so you had an opportunity to know this rule).

---

[55] Would you rather say that the second term is −2? We could say that the two terms $3x$ and positive 2 are subtracted to make $3x - 2$, or we could alternatively say that the two terms $3x$ and negative 2 are added together, $3x + (-2) = 3x - 2$. There really is no difference, unless you want to dive into the technical language of what exactly do we mean by a term.

If you know the rule that $x^{1/2} = \sqrt{x}$ from algebra,[56] you can find the derivative of $f(x) = \sqrt{x}$ using the formula that we learned for polynomial terms. With this rule, the given function is $f(x) = x^{1/2}$. Compare $x^{1/2}$ (which is equivalent to $1x^{1/2}$) with the general form ax^b to see that a = 1 and b = 1/2. The formula for the derivative of a polynomial term gives us $abx^{b-1} = (1)(1/2) \, x^{1/2-1} = (1/2) \, x^{-1/2}$. Using the rule from algebra that $x^{-p} = \dfrac{1}{x^p}$, this becomes $\dfrac{1}{2x^{1/2}}$. Since $x^{1/2} = \sqrt{x}$, the derivative of f with respect to x is equal to $\dfrac{1}{2x^{1/2}}$. Since you can't take the square root of a negative number (and obtain a real result[57]), this function and its derivative are only valid if x is positive.[58]

$$\frac{d}{dx}\sqrt{x} = \frac{d}{dx}x^{1/2} = \frac{1}{2}x^{1/2-1} = \frac{1}{2}x^{-1/2} = \frac{1}{2x^{1/2}} = \frac{1}{2\sqrt{x}}$$

Not all derivatives can be found using the formula abx^{b-1}; this formula is only for the terms of polynomials (or similar terms with fractional exponents, like we saw in the previous paragraph). There are many other kinds of functions, such as composite functions like the square root of $(3x^2 - 4)$, which is the square root of a polynomial (rather than a simple square root like the previous paragraph), trig functions like sine or cosine, log functions, or exponential functions. We'll consider a few of these other kinds of functions now.

Calculus students use their knowledge of trigonometry to find the derivatives of trig functions. Since trigonometry isn't a prerequisite for this conceptual book, we'll just show you visually how one of these derivatives relates to the slopes of tangent lines. A graph of the sine function is shown below. The sine function is periodic. The term periodic refers to the repeating pattern that you see below. It repeats because adding 360 degrees (which equates to 2π radians, where π is the Greek letter pi) to any angle makes an equivalent angle. For example, 90 degrees, 450 degrees, and 810 degrees are all equivalent angles, since $450 - 90 = 360$ and $810 - 450 = 360$.

---

[56] Why does $x^{1/2} = \sqrt{x}$? It follows from the rule $x^m x^n = x^{m+n}$. According to this rule, $x^{1/2}x^{1/2} = x^1$, since $1/2 + 1/2 = 1$. Since $x^1 = x$, this means that $x^{1/2}x^{1/2} = x$. Now compare $x^{1/2}x^{1/2} = x$ with $(3)(3) = 9$. Since 3 times 3 is 9 (meaning that 3 squared is 9), it follows that the square root of 9 is 3. If we apply the similar logic to $x^{1/2}x^{1/2} = x$, we find that the square root of x is equal to $x^{1/2}$.

[57] If you're not afraid of imaginary numbers, then you can factor a negative one out of the square root and replace it with the imaginary number i. But calculus is already hard enough with real numbers, so we'll try to keep it real in this book.

[58] This function is also valid if x is zero, but the derivative has a problem at $x = 0$, since it would involve division by zero.

sin x

At 90°, the slope is zero.

At 360°, the slope is +.

At 0°, the slope is +.

180°

450°

0°   90°   270°   360°   540°   → x

At 180°, the slope is −.

At 270°, the slope is zero.

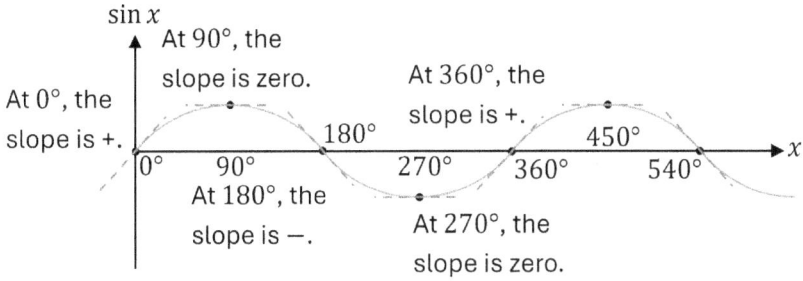

Examine the slopes of the sine function. The sine function above starts out with positive slope. The slope of the tangent lines decreases. At 90 degrees (which is $\pi/2$ radians), the slope of the tangent line is zero. From 90 degrees to 270 degrees, the slope is negative. At 180 degrees, the slope is negative and has maximum steepness (matching the steepness it had in the beginning, at 0 degrees, except that at 180 degrees, the slope is negative). At 270 degrees, the slope is again horizontal. The slope then becomes positive, reaching maximum positive steepness at 360 degrees. This pattern repeats every 360 degrees.

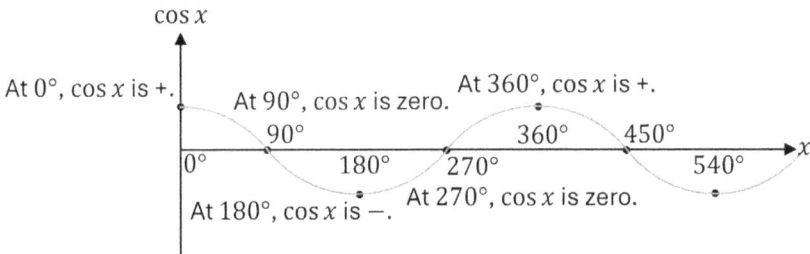

cos x

At 0°, cos x is +.

At 90°, cos x is zero.

At 360°, cos x is +.

90°

360°   450°

0°   180°   270°   540°   → x

At 180°, cos x is −.

At 270°, cos x is zero.

Now look at the cosine graph above and compare it with the sine graph. It turns out that the values of the cosine graph are equal to the slopes of the tangent lines of the sine graph. The cosine graph starts out at 1 at 0 degrees, matching the positive slope that the sine graph at 0 degrees. The cosine graph gets smaller from 0 to 90 degrees, just as the slope of the tangent lines of the sine graph become less steep. At 90 degrees, the value of cosine is zero, matching the horizontal slope of the sine graph. From 90 to 270 degrees, the cosine is negative, matching the negative slopes of the sine graph. At 180 degrees, the cosine equals negative 1, its most negative value, and the sine graph has its most negative slope at this point. At 270 degrees, cosine is zero, and the sine graph has a horizontal tangent line. From 270 to 360 degrees, the cosine graph rises, and the slopes of the tangent lines of the sine graph become steeper. This comparison demonstrates qualitatively that the cosine function gives the slopes of the sine graph. It's not just qualitative; it turns out to be exact numerically.[59] The cosine function is the derivative of the sine function. If y = sin(x), then dy/dx = cos(x). If you draw graphs of sine, cosine, tangent, and their reciprocals (for example, the secant function is the reciprocal of the cosine function), and if you study their slopes, you can similarly observe that the derivative of cosine is negative sine and that the derivative of tangent is one over cosine squared (an equivalent, yet more common, way to state this is to say that the derivative of tangent is secant squared).

The derivative of the exponential function e^x is very special in calculus. It turns out that this is the only function that happens to equal its own derivative. If y(x) = e^x, then $\frac{dy}{dx} = e^x$. What does this mean? It means that if you find the slope of the tangent line of a graph of e^x, the value of the slope of the tangent line happens to equal the same value as e^x. For example, when $x = 0$, $e^x = e^0 = 1$. The line tangent to e^x at $x = 0$ has a slope of 1. When $x = 2$, $e^x = e^2$ is approximately 7.389, and the line tangent to e^x at x = 2 has a slope of approximately 7.389. The agreement is exact, not approximate. When we said that e^2 is approximately 7.389, that's just because we rounded the calculator's answer to 3 places. The number e^2 has an infinite number of decimal places. Even so, the exponential function and its slope are identical for any value of x.

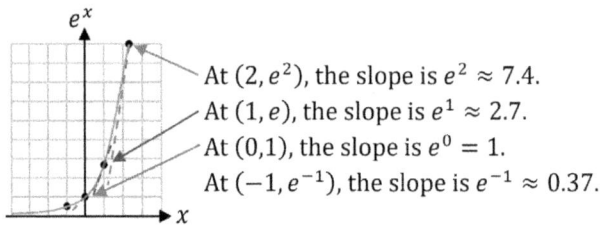

At $(2, e^2)$, the slope is $e^2 \approx 7.4$.
At $(1, e)$, the slope is $e^1 \approx 2.7$.
At $(0,1)$, the slope is $e^0 = 1$.
At $(-1, e^{-1})$, the slope is $e^{-1} \approx 0.37$.

In Chapter 5, we learned that exponential functions and logarithms are related. But while the exponential function is its own derivative, the derivative of the logarithm is considerably different from a logarithm. A derivative of ln(x), which is the natural logarithm of x (recall Chapter 5), with respect to x is equal to the reciprocal of x. That is, if y(x) = ln(x), then $\frac{dy}{dx} = 1/x$.

---

[59] Calculus students show that a derivative of sin(x) with respect to x is equal to cos(x) by using trig identities and the limits of ratios, such as the limit of sin(h)/h as h approaches zero (we'll explore this particular ratio in Chapter 13). That calculation is way beyond the mathematical prerequisites for this book, which is why we just demonstrated this visually (which is by no means a proof).

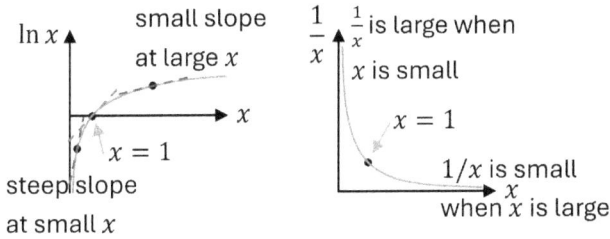

We will examine the relationship between $\ln(x)$ and its derivative, $1/x$, visually. See the graphs above. When x is small and positive, $1/x$ is large and positive, which agrees with the steep positive slope of $\ln(x)$. When x is one, $1/x$ is one and the slope of $\ln(x)$ equals one. As x grows larger, $1/x$ gets smaller and the positive slope of $\ln(x)$ becomes less steep. Comparing these graphs qualitatively, the values of $1/x$ agree with the behavior of the slope of $\ln(x)$.

Another important kind of function is a **composite** function. For example, consider $\sin(x^2)$, which means to first square x and then find the sine of x^2. Contrast this with $\sin^2(x)$, which means to first find $\sin(x)$ and then square the sine function. In $\sin(x^2)$, we square the angle but not the sine function, whereas in $\sin^2(x)$, we square the sine function but not the angle. Either case is an example of a composite function in that it combines two functions together: the sine function and a polynomial function.

The way to take a derivative of a composite function is to apply the **chain rule**. Suppose that f is a function of u and that u itself is a function of x. These two functions can be combined together to express f as a function of x. To take a derivative of f with respect to x, the chain rule says to take a derivative of with respect to u and multiply this by a derivative of u with respect to x. That is, $\frac{df}{dx} = \frac{df}{du}\frac{du}{dx}$. This is easier to understand by example.

As an example, consider the functions $f(u) = \sin(u)$ and $u(x) = x^2$. We may combine these two functions together to make a single composite function by replacing u with x^2 in the equation for $f(u)$. This makes the composite function $f(x) = f(u(x)) = \sin(x^2)$. The chain rule says that in order to take a derivative of f with respect to x (that is, in order to find $\frac{df}{dx}$), we should multiply $\frac{df}{du}$ by $\frac{du}{dx}$. Since $f(u) = \sin(u)$, from earlier in the chapter we know that $\frac{df}{du} = \cos(u)$. Since $u(x) = x^2$, from earlier in the chapter we also know that $\frac{du}{dx} = 2x$. When we multiply $\frac{df}{du} = \cos(u)$ by $\frac{du}{dx} = 2x$, we get $\frac{df}{dx} = \frac{df}{du}\frac{du}{dx} = \cos(u)$ times $2x = 2x\cos(u)$. In the last step, we used the commutative property of multiplication, meaning that the order in which two numbers is multiplied doesn't matter. We will replace u with x^2 to write the final answer as $df/dx = 2x\cos(x^2)$. If this example just looks Greek to you, which it does to many students who see this for the first time, consider that in this book, the main idea is that the chain rule provides a way of finding a derivative of a function that combines two different functions together. That's what you really need to know about the chain rule.

Thus far, we've used the notation dy/dx for a derivative of y with respect to x. This is one of two kinds of notations for derivatives that is commonly used. The other common notation uses a prime (') to indicate a derivative. That is, many books write y' (read as "y prime") instead of dy/dx. Each notation has benefits and drawbacks.

The prime notation is more concise. It's simpler to write y' than it is to write dy/dx. The prime notation is also handy to show that a quantity is a function and which variable it is a function of. For example, when we write y' for a derivative of y with respect to x, it's easy to write y'(x), which stresses that y is a function of x. If you do the same with dy/dx notation, it would look like dy(x)/dx or d/dx y(x), which is more cluttered.

The notation dy/dx makes it very clear which variable the derivative is taken with respect to. This notation is especially useful in the context of the chain rule, where one derivative is taken with respect to one variable (u) and another derivative is taken with respect to a different variable (x). It's easy to tell df/dx and df/du apart when we use this notation. With the prime notation, it's possible to express the chain rule as f'(x) = f'(u) u'(x), where f'(x) means df/dx while f'(u) means df/du, but this form of the chain rule isn't as common as $\frac{df}{dx} = \frac{df}{du}\frac{du}{dx}$ because the df/dx form is clearer. Since each notation has both advantages and disadvantages in different contexts, both notations are widely used, and it's common for the same textbook or instructor to use different notations in different contexts (to enjoy the benefits of each without the disadvantages). Calculus students really need to learn it both ways.

So, what was the main idea of this chapter? In calculus, students learn techniques for finding derivatives of functions. The derivative of a function tells us the slope of the tangent line at a particular point. As we'll explore in the next chapter, a derivative may alternatively be interpreted as the instantaneous rate of change of a quantity.

Try It Yourself (Ch. 8)

There are a few straightforward exercises below designed to help you feel like you're doing some real calculus. Don't be afraid; there is a lifeboat. If you find yourself drowning in the exercises, you can find the lifeboat at the end of this chapter, after "A Funny Thing about Logarithms" and just before Chapter 9. What is a lifeboat? It's the full solution to each exercise with explanations. But you can probably solve the exercises by yourself just by studying the example, provided that you can muster enough confidence. Believe in yourself.

Example. Given $y(x) = 5x^4$, find dy/dx. Compare ax^b with $5x^4$ to identify a = 5 and b = 4. Plug these into the formula abx^{b-1} to get dy/dx = $(5)(4)x^{4-1} = 20x^3$.

1. Given $y(x) = 7x^3$, find dy/dx.

2. Given $f(x) = x^6$, find df/dx.

3. Given $g(x) = 6x^7 - 9x^4 + 8$, find dg/dx.

A funny thing about logarithms (Challenge)

For most functions, if you multiply the argument by a constant, this has a noticeable effect on the derivative of the function. For example, consider the sine function $f(x)$ = $\sin(x)$, for which a derivative with respect to x equals df/dx = $\cos(x)$. If we multiply the argument of the function by the constant k, the function becomes $f(kx) = \sin(kx)$. To find df/dx, we should use the chain rule. Let u = kx. Then our function $f(kx)$ = $\sin(kx)$ takes the form $f(u) = \sin(u)$. To find df/dx, the chain rules says to multiply df/du with du/dx. Since $f(u) = \sin(u)$, df/du = $\cos(u)$, and since u = kx (and k is a constant like 2 or 5), du/dx = k. The chain rule then tells us that df/dx = $\cos(u)$ times k = k $\cos(u)$. Since u = kx, our final answer is df/dx = k $\cos(kx)$. Just look at the start and finish to see what happened. We began with $f(kx) = \sin(kx)$, and found that df/dx = k $\cos(kx)$. Compare that to when we simply had $f(x) = \sin(x)$, for which df/dx = $\cos(x)$. What changed when we multiplied the argument of the function by the constant k? Not only does this k show up in the argument of cosine, but we gained a coefficient k. The original answer, $\cos(x)$, became k $\cos(kx)$. When taking derivatives of functions where the argument of the function includes a factor of k similar to this, it's common for the k to show up in the derivative. Calculus students are accustomed to this. But when you try the same thing with the **logarithm** function, something 'funny' happens. Calculus students who don't pay close attention often overlook the 'funny' thing that happens (and get an incorrect answer when working with a logarithm that includes a constant in the argument).

Try it and see. Given $f(kx) = \ln(kx)$, where k is a constant, find df/dx. If you do it right, something 'funny' should happen in comparison to what happened in the previous paragraph with $\sin(kx)$. You can find the solution to this problem after the other solutions below.

Solutions to the Ch. 8 Exercises

1. Compare ax^b with $7x^3$ to identify a = 7 and b = 3. Plug these into the formula abx^{b-1} to get $\frac{dy}{dx} = (7)(3)x^{3-1} = 21x^2$.

2. Compare ax^b with x^6 to identify a = 1 and b = 6. (Recall from one of the examples in this chapter that the 'trick' here is to know that $1x^6 = x^6$.) Plug these into the formula abx^{b-1} to get $\frac{df}{dx} = (1)(6)x^{6-1} = 6x^5$.

3. Compare ax^b with $6x^7$ to identify a = 6 and b = 7 for the first term, which gives $(6)(7)x^{7-1} = 42x^6$. Compare ax^b with $9x^4$ to identify a = 9 and b = 4 for the second term (disregarding the minus sign for now; don't worry, we'll include the minus sign shortly), which gives $(9)(4)x^{4-1} = 36x^3$. The derivative of 8 with respect to x is zero because the derivative of any constant is zero (as discussed in an example in this chapter). Putting all of this together, $\frac{dg}{dx} = 42x^6 - 36x^3$.

Logarithm challenge. We will solve this problem two different ways. In our first solution, the only thing you need to know about logarithms is that d/dx of $\ln(x)$ equals $1/x$, as mentioned in the chapter. Following the example with the sine function, let u = kx. Then f(kx) = $\ln(kx)$ becomes f(u) = $\ln(u)$. Find df/du = 1/u and du/dx = k. The chain rule says to multiply these to get df/dx = (1/u) times k = k/u. Since u = kx, this becomes df/dx = k/(kx) = 1/x. The k cancels out. The 'funny' thing about a derivative of $\ln(kx)$ with respect to x is that the constant k cancels out. Compare d/dx of $\ln(x)$, which equals $1/x$, with d/dx of $\ln(kx)$, which also equals $1/x$. The constant k has no effect on the derivative. The 'funny' thing about $\ln(x)$ is that $\ln(kx)$ and $\ln(x)$ have exactly the same slope (for any value of x). Unlike the sin(kx) example (and other kinds of functions) where the k showed up in the derivative, the k cancels out in the derivative of $\ln(kx)$ with respect to x.

If you know about logarithms, there is a simple way to get the same answer. We will use the identity $\ln(kx) = \ln(k) + \ln(x)$. Now when we find df/dx, we get $\frac{d}{dx}\ln(k) + \frac{d}{dx}\ln(x)$ $= 0 + \frac{1}{x} = \frac{1}{x}$. The first term is zero because $\ln(k)$ is a constant. Since $\ln(kx) = \ln(k) + \ln(x)$, we see that $\ln(kx)$ is a constant plus $\ln(x)$. The effect of k in $\ln(kx)$ is just to add a constant to $\ln(x)$. The k doesn't have any effect on the slope of the logarithm because it simply raises the entire graph up[60] by a constant.

---

[60] Or down. If the constant k is less than Euler's number (e is approximately 2.71828), then $\ln(k)$ is negative, which shifts the graph of the logarithm down instead of up.

9 Which physical quantities involve derivatives?

You may recall that we discussed measuring distance and time for a car traveling along a straight road back in Chapter 3. We'll consider a similar example now.

Suppose that a car is traveling along a straight road. The car's speed is changing non-uniformly, and we wish to determine the car's speed at a particular point on the road. The speed at a particular instant (or position) is called the **instantaneous** speed. Suppose that we have electronic sensors and software that record the position of the car every millisecond. Since we have exactly one value for the position of the car (which we'll call x) for each value of time (which we'll call t), the position meets the definition of a function from Chapter 5. In particular, position is a function of time: $x(t)$.

If we plot x as a function of t, a derivative of position with respect to time tells us the slope of the tangent line. Something may seem strange about the notation here. In Chapter 8, with $y(x)$, we made a graph with y on the vertical axis and x on the horizontal axis, which is the standard coordinate system that students learn in algebra. Now with our car problem, we have $x(t)$. Compare $y(x)$ with $x(t)$. Note that x is in a different place in this notation. In the car problem, x goes on the vertical axis and time goes on the horizontal axis. In the physics of motion, we treat the time (t) as the independent variable and the dependent variable (x) is a function of time: $x(t)$. Suddenly, in physics, x appears on the vertical axis (where y usually is), and instead t appears on the horizontal axis. It may seem strange at first, but it makes sense from the perspective that in motion, position (x) is the dependent variable and time (t) is the independent variable; that is, position (x) is a function of time (t). Students need to be able to adapt. (If you want to understand calculus, you need to be able to change your mindset sometimes. It would be silly to think you could understand calculus, which is the mathematics of change, without ever changing the way you think.)

In the previous chapter, $y_2 - y_1$ was the rise, $x_2 - x_1$ was the run, and we took the limit as $x_2 - x_1$ went to zero to find the slope of the tangent line, which gave us the derivative of y with respect to x, dy/dx. For the moving car, the rise is $x_2 - x_1$, the run is $t_2 - t_1$, and we want the limit as $t_2 - t_1$ goes to zero. By comparison, the slope of the tangent line of the graph (of x as a function of t) for the car's motion is a derivative of x with respect to t, $\frac{dx}{dt}$. Now let's explore what this derivative, dx/dt, means.

If you can recall the car example from Chapter 3, we were thinking about dividing the distance traveled by the time taken for a short time interval. On our graph, $x_2 - x_1$ is the distance traveled when the time interval is $t_2 - t_1$. Taking the limit as the time interval approaches zero of the ratio of $x_2 - x_1$ to $t_2 - t_1$ is exactly the same as finding the slope of the tangent line of a graph of x as a function of t. When we think about dividing the distance by the corresponding time interval in the limit that the time interval becomes zero, we're finding the instantaneous speed of the car, much as we had discussed in an example in Chapter 3. When we divide the distance $x_2 - x_1$ by $t_2 - t_1$, we're finding the slope that joins two points, and when we take the limit that the time interval $t_2 - t_1$ approaches zero, this slope becomes the slope of the tangent line. Thus, we see that the calculation that we had considered back in Chapter 3, dividing the distance by the corresponding time interval in the limit that the time interval approaches zero, is identical[61] to the calculation that we did in the previous chapter to find the slope of the tangent line. Therefore, a derivative of position with respect to time, dx/dt, tells us the instantaneous speed of the car. The two ways of thinking about a derivative (the slope of the tangent line when we plot x as a function of time, or the instantaneous rate of change of position when we divide the distance traveled by the time interval in the limit that the time interval approaches zero) are equivalent.

Technically, **velocity** is a derivative of position with respect to time. Velocity is the instantaneous rate of change of position. The words speed and velocity are closely related, but they are different. The distinction is that velocity is a combination of speed and direction. The **speed** of an object tells you how fast it is moving, whereas **velocity** tells you both how fast it is moving and which way it is headed. For an object traveling in a straight line, velocity can be positive or negative because the object could be moving forward or backward, whereas the speed is the absolute value of velocity. (For an object traveling in two or three dimensions, instead of traveling in a straight line, the relationship is more complicated.[62]) In straight-line motion (using x for the position of an object along this line) the velocity is $\frac{dx}{dt}$ and the speed is $|\frac{dx}{dt}|$.[63]

---

[61] Identical, except for having x on the vertical axis and t on the horizontal axis.

[62] In a physics course, students learn about vectors (which, like velocity, include direction) and scalars (which, like speed, don't have direction). The magnitude of the velocity vector equals the speed. Given the components of the velocity vector, students use the Pythagorean theorem to find the magnitude of the velocity, which is the speed.

[63] We sometimes call these the instantaneous velocity and instantaneous speed to distinguish them from average velocity and average speed, but the adjective 'instantaneous' is often omitted. If no adjective is used, instantaneous is implied. If you want to refer to an average, then you have to include the word 'average.'

If you take a derivative of velocity with respect to time, you get acceleration. The **acceleration** tells you how the velocity changes. There are three ways that an object can change velocity: it can speed up, it can slow down, or it can change direction (like a car turning left). If a car travels in a straight line with an acceleration of 5 m/s² (meters per second squared), this means that every second, the car is moving 5 m/s faster than it was the previous second. You can think of 5 m/s² as 5 (m/s)/s, meaning that it gains 5 m/s of speed every second. If a car travels in a straight line with an acceleration of −5 m/s², this means that the car loses 5 m/s of speed each second.[64] Since velocity is a derivative of position with respect to time and since acceleration is a derivative of velocity with respect to time, this means that acceleration is the **second derivative** of position with respect to time: $\frac{d^2x}{dt^2}$. The notation $\frac{d^2x}{dt^2}$ means to first find the derivative $\frac{dx}{dt}$ to obtain velocity, and then take a derivative of the derivative to obtain acceleration.

Acceleration is a very important derivative in physics because of Newton's second law of motion. According to **Newton's second law**, the net external force acting on an object equals the object's mass times its acceleration: F = ma. Unfortunately, most students who learn physics don't get to appreciate that there is calculus involved in this seemingly simple equation.[65] That's because acceleration is often constant or zero. In more advanced courses, where there may be non-uniform acceleration, students replace a with $\frac{d^2x}{dt^2}$ in the right-hand side of F = ma, which makes a second-order differential equation. (Solving a differential equation is beyond the scope of this book; differential equations is a course that math, physics, and engineering students take after they've already passed a few semesters of calculus.) Once they solve the differential equation, this solution lets them calculate the position or velocity of the object at any given time.[66]

---

[64] If you want to get technical, if the car is heading in reverse, it alternatively could mean that the car is moving 5 m/s faster in the negative direction each second.

[65] Physics students generally don't think that this equation is simple. Although F = ma may look simple, since F is really the vector sum of multiple forces, students have to think about the different forces acting on an object and then use trigonometry to find the components of those forces. But the reality is that the equation is even more complicated than that when you consider that the acceleration is the second derivative of position with respect to time.

[66] Actually, it turns out that there are alternatives to using Newton's second law to do this. There are other ways to set up a differential equation to find the position or velocity at any given time. Two popular methods are Lagrange's equation and Hamilton's equation. These equations are equivalent to Newton's second law, but formulate the problem by considering the energy of the system rather than the forces acting on the object. One way that working with energy is

Gravitational acceleration is one kind of acceleration that every human has at least limited experience with. We have some sense for how objects travel if you throw them in the air, and what happens to our bodies if we jump or fall through a short distance (that is, short compared to the thickness of earth's atmosphere). If the only force acting on an object is the pull of gravity, the object experiences **gravitational acceleration**. Such an object is said to be in **free fall** (even if it happens to be momentarily headed upward[67]). Any projectile (which is an object traveling through the air) approximately experiences gravitational acceleration if the effects of air resistance are negligible. Near earth's surface, gravitational acceleration is 9.8 m/s^2. As an example, if you throw a rock straight upward, the rock loses nearly 10 m/s of speed each second on the way up and gains nearly 10 m/s of speed each second on the way back down. If the rock has an initial speed of 30 m/s, it will be traveling 20.2 m/s after one second, 10.4 m/s after two seconds, 0.6 m/s after three seconds, out of speed after 3.06 seconds (when it is momentarily at the top of its trajectory), 9.2 m/s and headed downward after four seconds, 19.0 m/s and still headed downward after five seconds, 28.8 m/s and still headed downward after six seconds, and the rock will return to its starting point 6.12 seconds after it leaves your hand.[68]

It turns out that many physical quantities are derivatives with respect to time. These include power (a derivative of work with respect to time), electric current (a derivative of charge with respect to time), and just about any kind of rate, such as a reaction rate (a derivative of the concentration of reactants in a chemical reaction with

---

simpler than working with force is that energy is a scalar (it doesn't have direction), whereas force is a vector (so we use trigonometry to resolve the forces into components).

[67] If you throw a rock upward, it will lose speed on the way up until it eventually runs out of speed, at which point it will gain speed coming back down. On the way up, on the way down, and even for the instant at the top, the rock experiences gravitational acceleration.

[68] These calculations assume that the effects of air resistance are negligible. That's not quite true, but air resistance is only a slight effect for many rocks at these speeds. Neglecting air resistance makes it easier to see what gravitational acceleration means. We subtracted 9.8 m/s from the speed each second going up (that's how we got 30 − 9.8 = 20.2 m/s, 20.2 − 9.8 = 10.4 m/s, and 10.4 − 9.8 = 0.6 m/s) and added 9.8 m/s to the speed each second going down (that's how we got 9.2 + 9.8 = 19.0 m/s and 19 + 9.8 = 28.8 m/s). To find the time it takes to reach the top of the trajectory, divide 30 m/s by 9.8 m/s^2 to get 3.06 seconds. Double this for the round trip. How did we get the number 9.2 m/s? We took the velocity at 0.6 m/s, which is positive since the object was still traveling upward after 3 seconds), and subtracted 9.8 m/s to get negative 9.2 m/s for the velocity after 4 seconds, then disregarded the minus sign to find speed (whereas velocity includes direction, speed does not).

respect to time), precipitation rate (a derivative of rainfall or snowfall with respect to time), or population growth rate (a derivative of the population with respect to time).

But many physical quantities involve derivatives with respect to some variable other than time. Examples include linear charge density (a derivative of charge with respect to position[69]), heat capacity (a derivative of heat with respect to temperature), and marginal revenue (the derivative of a company's revenue with respect to the amount of product manufactured).

Pop Quiz (Ch. 9)

See if you remember these points from Chapter 9.

1. If you take a derivative of position with respect to time, what do you get?

2. If you take a derivative of velocity with respect to time, what do you get?

10 What are extrema?

As it turns out, the slope of the tangent line of a function can tell us a lot about the function. Since the derivative of a function gives us the slope of the tangent line, this means that the derivative of a function can tell us a lot about the function itself. We'll see examples in this chapter of what we can learn about a function by analyzing its derivative.

Consider the function $y(x)$ that appears in the graph below. We divided the curve into three regions (labeled I, II, and III). Region I corresponds to $x < A$, Region II corresponds to $A < x < B$, and Region III corresponds to $B < x$. Consider the slope of the tangent line in each region.

---

[69] If the distribution of charge happens to be uniform, as is the case in many straightforward physics problems, in that special case, no derivative is needed. For a uniform charge density, simply divide the total charge by the length of the object. For the more general case where the charge is distributed non-uniformly, then a derivative is necessary. Physics students don't see this case as often because the math is more tedious, but non-uniform distributions of charge do occur in nature. For example, if a copper rod with more electrons than protons (so it has a net negative charge) is placed near (without touching) a positively charged object, the electrons on the surface of the copper rod will redistribute with a greater density of electrons near the positive object and fewer on the other end of the rod, forming a non-uniform distribution of charge. If you remove the positively charged object, then the electrons on the rod will distribute themselves uniformly (that is, with even spacing between electrons).

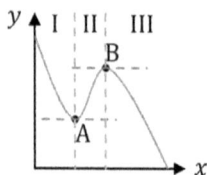

The tangent line is horizontal at the relative minimum (A) and relative maximum (B).

- In Region I (x < A), the slope of the tangent line is negative. The values of y are decreasing in Region I.
- At x = A, where Regions I and II meet, the tangent line is horizontal. The slope of the tangent line is zero at x = A. The value of y is a local minimum at x = A.
- In Region II (A < x < B), the slope of the tangent line is positive. The values of y are increasing in Region II.
- At x = B, where Regions II and III meet, the tangent line is horizontal. The slope of the tangent line is zero at x = B. The value of y is a local maximum at x = B.
- In Region III (B < x), the slope of the tangent line is negative. The values of y are decreasing in Region III.

The graph above illustrates a few important properties that relate to the slope of the tangent line. Since the derivative $\frac{dy}{dx}$ gives us the slope of the tangent line of y(x), without even looking at a graph, we can use the derivative $\frac{dy}{dx}$ to determine these very same properties.

- When the slope of the tangent line is positive (like Region II above), the values of y are increasing. That is, when $\frac{dy}{dx}$ > 0, the values of y are increasing.
- When the slope of the tangent line is negative (like Region I or III above), the values of y are decreasing. That is, when $\frac{dy}{dx}$ < 0, the values of y are decreasing.
- When the tangent line is horizontal (meaning that the slope of the tangent line is zero, like it is at x = A and x = B above), the value of y is a local minimum, local maximum, or a point of inflection (which we will define shortly). That is, when $\frac{dy}{dx}$ = 0, the value of y is a local minimum, local maximum, or a point of inflection.

The last bullet point above tells us how to use calculus to find the extreme values of a function. The **extreme values** of a function include local minima (sometimes called relative minima) and local maxima (sometimes called relative maxima). Local minima and local maxima occur when the tangent line is horizontal, which is where the slope of the tangent line is zero, which is also where the derivative of the function is zero. The points where the derivative is zero are called **critical points**.

Given a function $y(x)$ that is smooth and continuous, without even looking at a graph, one way to find all of the extreme values of a function is to find the derivative dy/dx and set the derivative equal to zero: $dy/dx = 0$. When we solve the equation that we get by setting dy/dx equal to zero, this gives us the values of x corresponding to **critical points**: local minima, local maxima, and points of inflection (since these are the points on a smooth curve where the tangent line is horizontal). The figure below shows a local minimum at x = A, a local maximum at x = C, and a point of inflection at x = B. At the local minimum (at x = A), the slope (and hence dy/dx) just before A is negative and the slope just after A is positive. At the local maximum (at x = C), the slope (and hence dy/dx) just before C is positive and the slope just after C is negative. For a local extremum (meaning minimum or maximum), the slope just after the point has the opposite sign compared to the slope just before the point. This isn't the case for a **point of inflection**, where the slope has the same sign before and after. At the **point of inflection** below (at x = B), the slope (and hence dy/dx) just before B and just after B are both positive.

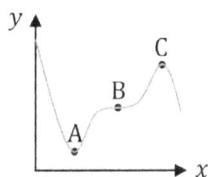

There is a local minimum at A, a point of inflection at B, and a local maximum at C.

The following **first derivative test** can be used to find extreme values and points of inflection,[70] given a function $y(x)$ that is smooth and which doesn't have any discontinuities:

• Take a derivative of the function with respect to the independent variable: $\frac{dy}{dx}$.
• Set the first derivative equal to zero: $\frac{dy}{dx} = 0$.
• Solve the resulting equation for x. Each value of x that solves the equation corresponds to a local minimum, a local maximum, or a point of inflection.
• Use the given function $y(x)$ to find the value of y corresponding to each value of x.
• For each critical point that you found, evaluate the first derivative just before and just after that value of x. If the sign of dy/dx changes from negative to

---

[70] This will find points of inflection where the slope of the tangent line is horizontal. If there happens to be a point of inflection with a vertical slope, the first derivative test as it is described below won't find that.

positive, it's a local minimum, if the sign of dy/dx changes from positive to negative, it's a local maximum, and if dy/dx doesn't change sign, it's a point of inflection.

For example, consider the function $y(x) = 3x^2 - 12x + 16$. We will apply the above strategy to find its extreme values.

- To find the derivative of polynomial terms, use the formula abx^{b-1}. Compare ax^b with $3x^2$ to see that a = 3 and b = 2 for the first term, which gives $(3)(2)x^{3-2} =$ $6x^1 = 6x$. Compare $12x$ with ax^b to see that a = 12 and b = 1 (since $x^1 = x$) for the middle term, which gives $(12)(1)x^{1-1} = 12x^0 = 12(1) = 12$ (since $x^0 = 1$). Most calculus students look at $12x$ and just know that the derivative with respect to x is 12 immediately (since $y = 12x$ is a straight line with a slope of 12). The last term is constant and the derivative of a constant is zero, so the last term won't make a contribution to the derivative. Putting all of this together (which most calculus do in a single quick step), we get $\dfrac{dy}{dx} = 6x - 12$.
- Set the first derivative equal to zero: $6x - 12 = 0$.
- Solve the above equation. Add 12 to both sides to get $6x = 12$. Divide by 6 on both sides to get $x = 12/6 = 2$. This problem just has a single solution: $x = 2$.
- Plug $x = 2$ into the equation $y(x) = 3x^2 - 12x + 16$ to find the corresponding value of y: $y(2) = 3(2)^2 - 12(2) + 16 = 3(4) - 24 + 16 = 12 - 8 = 4$. Since $x = 2$ and y = 4, the (x, y) coordinates of the point are $(2, 4)$.
- The derivative is $\dfrac{dy}{dx} = 6x - 12$. Evaluate the derivative at $x = 1.9$, which is slightly smaller than $x = 2$. This gives dy/dx = 6(1.9) – 12 = –0.6, which is negative. Now evaluate the derivative at $x = 2.1$, which is slightly larger than x = 2. This gives dy/dx = 6(2.1) – 12 = 0.6, which is positive. Since the derivative changed sign from negative to positive,[71] the point $(2, 4)$ is a local minimum.

---

[71] One problem with this first derivative test is that you need to be careful choosing the values of x just before and just after each critical point. We found a critical point at $x = 2$. We found that the derivative was negative at $x = 1.9$, which is slightly less than 2. What if the derivative had been negative at $x = 1.9$ and then turned positive at $x = 1.95$? That's not the case in this example, but it could conceivably happen in some other example. (However, if the function is smooth and continuous, there should in that case be another critical point between $x = 1.9$ and $x = 1.95$, so it shouldn't come as a surprise. But since some common functions have discontinuities or non-smooth points, mathematicians, scientists, and engineers need to be aware of the limitations of the first derivative test. The second derivative test discussed in the

The minimum value of y for this function is 4. You can see this visually in the graph below.

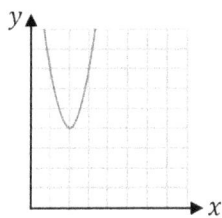

The parabola $3x^2 - 12x + 16$ has a minimum at $(2, 4)$, which is 2 units to the right and 4 units above the origin. At this point, the derivative $\frac{dy}{dx} = 6x - 12$ equals zero.

Note that if y goes to plus or minus infinity at any value of x, the first derivative test won't identify these values. For example, consider the graph of $y(x) = 1/x$ below. This graph is a hyperbola. It has two asymptotes (which are lines that the function approaches, but never quite reaches): one is the x-axis and the other is the y-axis. As x approaches zero from the right, y goes to positive infinity, and as x approaches zero from the left, y goes to negative infinity. (Does this ring a bell? If you remember what we learned about limits in previous chapters, you should know that the limit of this function does not exist at the value $x = 0$.)

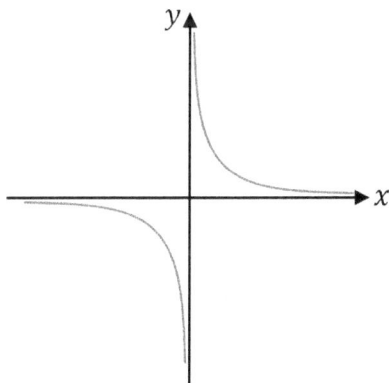

If $y(x)$ isn't smooth or if $y(x)$ has any discontinuities (sudden jumps), there may be local extrema where the tangent line isn't horizontal. For example, consider the graph of $y(x) = |x|$ below, where the vertical lines indicate absolute values (which mean to disregard any negative signs). This graph makes a V-shaped angle at $x = 0$, which is an abrupt change of direction rather than a smooth curve. Although there is

next chapter is arguably better, and is more commonly used, than the first derivative test discussed in this chapter.)

a local minimum at $x = 0$, the first derivative (and tangent line) is undefined at this point.

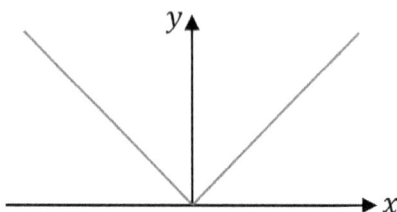

In calculus, students sometimes search for absolute extrema, rather than local extrema. If a function is specified over a finite interval, starting at $x = A$ and ending at $x = F$, one of the endpoints could be the absolute minimum or maximum. If an endpoint is an absolute extremum, the first derivative (and slope of the tangent line) may not be zero there. For example, in the graph below, y is higher at $x = F$ than the local maximum at $x = B$. Although there is a **local** maximum at $x = B$, the **absolute** maximum occurs at $x = F$. In contrast, the local minimum at $x = C$ is the absolute minimum because no other value of y is smaller than the value of y at $x = C$. When calculus students proceed to find absolute extrema, they first find local extrema and then compare them with the endpoints.

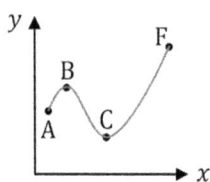

B is a local maximum, but F is the absolute maximum over the interval shown. Observe that $\frac{dy}{dx}$ is zero at B and C, but not at F.

Fill in the Blank (Ch. 10)

Try to fill in each blank below with the best word.

1. The minimum or maximum value of a function is a(n) _____ value.

2. At a local minimum or maximum of a smooth, continuous function, the first derivative is equal to _____.

11 What is concavity?

The technique that calculus students are taught for finding the extreme values of a function involves finding both the first and second derivatives of the function. In the previous chapter, we focused solely on the first derivative. In this chapter, we'll discover how the second derivative can help.

Given a continuous function $y(x)$, the first derivative dy/dx tells us the slope of the tangent line. A local minimum is formed when dy/dx is negative and then turns positive. A local maximum is formed when dy/dx is positive and then turns negative. These two cases are illustrated below. In either case, $dy/dx = 0$ at the extremum, where the tangent line is horizontal.

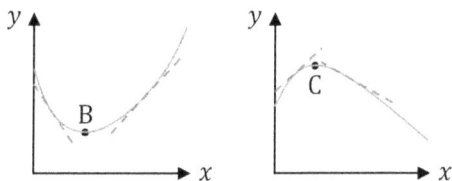

The distinction between a local minimum (B) and a local maximum (C) is whether the slopes of the tangent lines turn from negative to positive, or vice-versa.

The second derivative d^2y/dx^2 (which means to take a derivative of the derivative) tells us about the slope of the first derivative dy/dx. It turns out that if the function $y(x)$ has a local minimum or a local maximum, this relates to the sign of the second derivative as follows:

• Suppose that $y(x)$ has a local minimum at $x =$ B. In that case, dy/dx is negative just before $x =$ B, dy/dx is zero at $x =$ B, and dy/dx is positive just after $x =$ B, as shown in the left figure above. This means that the first derivative dy/dx is increasing at $x =$ B (since it goes from negative to zero to positive). If we find the second derivative d^2y/dx^2 and evaluate it at $x =$ B, the second derivative will be positive. Why? Because the slope of dy/dx is positive, since dy/dx rises from negative to positive.

• Suppose that $y(x)$ has a local maximum at $x =$ C. In that case, dy/dx is positive just before $x =$ C, dy/dx is zero at $x =$ C, and dy/dx is negative just after $x =$ C, as shown in the right figure above. This means that the first derivative dy/dx is decreasing at $x =$ C (since it goes from positive to zero to negative). If we find the second derivative d^2y/dx^2 and evaluate it at $x =$ C, the second derivative will be negative. Why? Because the slope of dy/dx is negative, since dy/dx falls from positive to negative.

According to the two bullets points above, the second derivative tells us about the concavity of the function at points B or C. **Concavity** refers to the shape of the curve. More specifically, we use the terms concave up and concave down to refer to the shape of a curve as follows:

- A curve $y(x)$ is **concave up** in a region where its first derivative dy/dx increases. In such a region, the slopes of the tangent lines are increasing. (If the slopes are positive, then they are getting steeper, meaning more positive. If the slopes are negative, they are getting less steep, meaning less negative.) The tangent lines are turning counterclockwise in such a region. The graph below on the left is concave up. The second derivative d^2y/dx^2 is positive if $y(x)$ is concave up.
- A curve $y(x)$ is **concave down** in a region where its first derivative dy/dx decreases. In such a region, the slopes of the tangent lines are decreasing. (If the slopes are positive, then they are getting less steep, meaning less positive. If the slopes are negative, they are getting steeper, meaning more negative.) The tangent lines are turning clockwise in such a region. The graph below on the right is concave down. The second derivative d^2y/dx^2 is negative if $y(x)$ is concave down.

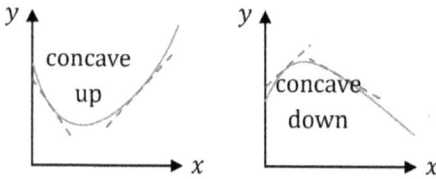

Concave up: shaped like the letter U, the tangent lines are gaining slope (they turn counterclockwise). Concave down: the tangent lines are losing slope (they turn clockwise).

At a local minimum, $y(x)$ is concave up (left figure above). At a local maximum, $y(x)$ is concave down (right figure above). It's easy to forget which is concave up and which is concave down. Note that when a curve is concave up, the curve is shaped like the letter U (left figure above). Another way to help keep these two terms straight is to examine the sign of the second derivative: $y(x)$ is concave up where the second derivative is positive, and $y(x)$ is concave down where the second derivative is negative. But beware: A point that sometimes causes confusion is that a local minimum occurs in a region that is concave up, while a local maximum occurs in a region that is concave down. When you think of concave 'up,' think of the letter U, which has a minimum, and that may help to avoid this confusion.

We can use the first and second derivatives together to search for local extrema as follows:

• Take a derivative of the function with respect to the independent variable: dy/dx.

• Set the first derivative equal to zero: $dy/dx = 0$.

• Solve the resulting equation for x. Each value of x that solves the equation corresponds to a local minimum, a local maximum, or a point of inflection.

• Use the given function $y(x)$ to find the value of y corresponding to each value of x. (So far, every step has been the same as the first derivative test from the previous chapter. The remaining steps will be different.)

• Take a derivative of dy/dx to find the second derivative d^2y/dx^2.

• For each critical point that you found, evaluate the second derivative.

• If the second derivative is **negative**, the curve is concave down at that value of x, which corresponds to a local **maximum**. If the second derivative is **positive**, the curve is concave up at that value of x, which corresponds to a local **minimum**. If the second derivative is **zero**, the test is inconclusive. It's possible that the point is a point of inflection, but not necessarily.[72] To find out, you can check the values of y when x is a little smaller or a little higher than the critical value, like we did in the previous chapter.

For example, consider the function $y = -x^2 + 10x + 7$.

• The first derivative is $dy/dx = -2x + 10$ and the second derivative is $d^2y/dx^2 = -2$.

• When we set the first derivative equal to zero, we get $-2x + 10 = 0$.

• Add $2x$ to both sides to get $10 = 2x$. Divide by 2 on both sides to get $5 = x$. There is one critical point at $x = 5$.

• Plug $x = 5$ into the given equation to find the corresponding value for y. We get $y = -x^2 + 10x + 7 = -5^2 + 10(5) + 7 = -25 + 50 + 7 = 25 + 7 = 32$. The (x, y)

---

[72] An obvious exception occurs if $y(x)$ is a straight line. For example, if $y = 2x + 7$, $dy/dx = 2$ is a constant such that d^2y/dx^2 is zero. Here, there is no concavity because the function isn't a curve – it's a straight line. But there are also more subtle exceptions. For example, for $y = x^4$, $dy/dx = 4x^3$ and $d^2y/dx^2 = 12x^2$. Setting $dy/dx = 0$, we see that $x = 0$ is a critical point. At $x = 0$, the second derivative is zero, which is inconclusive. In this case, $y = x^4$ has a local minimum at $x = 0$, even though the second derivative isn't positive here. But the first derivative $dy/dx = 4x^3$ is negative for $x < 0$, zero for $x = 0$, and positive for $x > 0$, which shows that dy/dx is increasing in this region, which shows that this is a local minimum. On the other hand, when d^2y/dx^2 is zero at a critical point, there often is a point of inflection. That's the case with $y = x^3$, for example. Here, $dy/dx = 3x^2$ and $d^2y/dx^2 = 6x$. There is critical point at $x = 0$, where $d^2y/dx^2 = 0$. In this case, the first derivative $dy/dx = 3x^2$ is positive if $x < 0$ and is also positive if $x > 0$, showing that this is a point of inflection.

coordinates of the critical point are (5, 32).

• Since the second derivative in this example happens to equal −2 for any value of x, we don't need to plug in $x = 5$ to find that the second derivative is negative at the critical point. Since d^2y/dx^2 is negative, there is a local maximum at $x = 5$. The maximum value of y is 32.

The second derivative test (which tests for the concavity of a function) has some important applications in physics and engineering, especially when a system is at or near equilibrium. A state of **equilibrium** is a balanced state that a system tends to attain in 'natural' conditions. For example, a rigid body is in static equilibrium if the net external forces and torques[73] are zero; in this example, 'natural' means the absence of net external forces or torques. As another example, when an ice cube is placed in a cup of warm liquid, thermodynamic equilibrium is attained once the ice and liquid both reach the same temperature (at which point the ice will have melted into liquid water). Many systems in nature tend towards equilibrium states. We will explore a couple of examples and consider how the second derivative test (which relates to the concavity of a graph of a function) relates to the equilibrium state.

As our first example, consider the curved surface illustrated below. It's a rigid surface, perhaps made of wood, concrete, or metal, so that it won't change shape when a small object is placed on it. Imagine that we wish to place a small ball at some point on this surface. Where could we place the ball so that it would be in equilibrium? The ball will be in equilibrium if the net external force acting on the ball is zero, but you don't really need to know physics to answer this question. If you have any experience with balls rolling on surfaces in real life, you may be able to determine the answer. But so that you don't feel like you're missing out on any key information, a force is a push or a pull. There are two forces to consider in this problem. One is the force of gravity, which we call the weight of the ball. The other is called normal force, and it's the force that the surface exerts on the ball. The normal force is a support force in that it has an upward component (supporting the object, at least in part, against gravity). The normal force is the reason the ball doesn't fall straight down (the way it would if there were no surface underneath it). The word 'normal' in physics and math means perpendicular. Here, the direction of the normal force is perpendicular to the surface. But again, common human experience should be enough to answer the question; knowledge of the underlying physics isn't crucial for this example.

---

[73] A force is a push or a pull, and a torque is exerted by a force when it affects the rotation of an object.

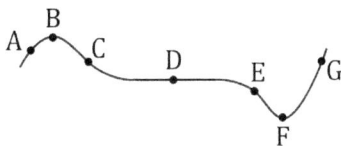

We'll approach the problem using physics and calculus, but to answer the first question above, you probably don't need any physics or calculus, as we'll eventually see. The physics involves potential energy, so let's quickly explain what it is. Energy is the ability to do work, and work is done when a force contributes toward the displacement of an object. Potential energy is work that can be done by changing position. In this problem, the ball has gravitational potential energy, which is work that can be done by gravity. Remove the surface and you'll see gravity do that work; the ball will fall downward. Gravitational potential energy depends on height. The higher the ball is, the more work that gravity can do bringing it down. We will use the symbol U for potential energy, and the formula[74] is U = mgy, where m is the mass of the ball, g is gravitational acceleration (about 9.8 m/s² near earth's surface), and y is the height of the ball relative to the origin (which we might place on the floor, for example).

Let the function $y(x)$ represent the curve corresponding to the surface above. Then U is a function of x also, $U(x)$ = mg times $y(x)$. We can phrase our question as, for which values of x is the ball in equilibrium? As we'll see shortly, we don't really need to use calculus to answer this first question. So why use calculus? For one, we'll ask a follow-up question that will help us see how the second derivative and concavity are involved. For another, some equilibrium problems are more abstract and harder to 'see' (as in our subsequent examples).

One more bit of physics is needed. For a conservative[75] force, the horizontal

---

[74] Students who have studied physics might know this formula as PE = mgh. We're using U instead of PE here because we plan to take derivatives. It's simpler to write dU/dx (yeah, we have y right now, not x, but be patient; we'll get there in the text) than it is to write dPE/dx, so using U will help reduce confusion. Many advanced physics texts (that is, those which will use ample calculus) prefer U to PE for this reason. And why are we using y instead of h? We'll be treating y as a function of another variable x, and since we've been using $y(x)$ already in this chapter, that may seem less daunting than introducing h. We're already throwing a lot of different symbols at readers who may have never seen these formulas before, so we're trying to keep it as friendly as possible.

[75] When a conservative force does work, one form of energy is transformed into another in a way that mechanical energy is conserved. For example, gravitational force is conservative. If a

component of the force is proportional to the negative of the first derivative of the potential energy:[76] $F = -dU/dx$.

Now we're ready to put the pieces of this puzzle together. We said that the ball would be in equilibrium if the net external force is zero. Set $F = 0$ to get $0 = -dU/dx$. Multiply by -1 on both sides to write this more simply as $0 = dU/dx$. We've basically seen this before. This says to set the first derivative of the potential energy equal to zero. We learned in the previous chapter that this helps us find the critical points. So, the ball is in equilibrium at the critical points. Since $U = mgy$, our equation becomes $0 = d/dx$ of mgy. Since m and g are constants, $0 = mg$ times dy/dx. Divide by mg on both sides: $0 = dy/dx$. (Basically, we're saying that if $dy/dx = 0$, multiplying by mg doesn't matter.) So, the critical points occur where $dy/dx = 0$, where $y(x)$ is the equation of the curve for the surface. If you've been paying attention in the previous chapter and this chapter, you should know that $dy/dx = 0$ at the local minimum, local maximum, and point of inflection. These are points F, B, and D in the figure above. But you could have guessed this without knowing any physics or calculus at all. If you place a marble on a table, under which circumstances would the marble stay on the table or roll off? If the table isn't level, the marble will roll off the table. Where is the above surface level? It's level where the tangent is horizontal, at points F, B, and D, which are the critical points.

Now for the follow-up question, where the second derivative and the concept of concavity will be applicable. At which points would the ball be in **stable equilibrium**? By stable, we mean if we give the ball a slight push, the ball will oscillate back and forth about equilibrium. If instead the ball would roll far away, that's not stable equilibrium. (You might be able to determine this answer too without using any knowledge of physics or calculus, but we want to use calculus here to see how the question is related to the second derivative.)

---

ball falls due to gravity, its gravitational potential energy is transformed into kinetic energy. Resistive forces like friction and air resistance, on the other hand, subtract from the total mechanical energy of the system. (If the ball loses mechanical energy due to air resistance or friction, this goes into other forms of energy, like the internal energy of the ball or heat energy transferred to the surface or to the air.)

[76] We're using F for the x-component of the force just to minimize how intimidating the notation might seem. In physics, we would write it using a subscript as F_x. Since we're only going to consider one component of the force (along x), we don't need the subscript to distinguish between F_x and F_y, so we'll just write F for simplicity.

It turns out that the sign of the second derivative at a point of equilibrium (a critical point) determines whether the equilibrium is stable or unstable:

• d^2U/dx^2 is positive at a point of **stable equilibrium**, where $U(x)$ and $y(x)$ are at a local minimum (which is concave up). In the previous diagram, a ball would be in stable equilibrium if placed at point F. If it is pushed slightly away from point F, the ball will roll back and forth about point F, in the bottom of the valley.

• d^2U/dx^2 is negative at a point of **unstable equilibrium**, where $U(x)$ and $y(x)$ are at a local maximum (which is concave down). In the previous diagram, point B is a position of unstable equilibrium. If it is pushed slightly away from point B, the ball will roll away from point B. The ball would need to be exactly at point B, perfectly balanced without any wind or vibrations, in order to remain there.

• If d^2U/dx^2 is zero, more information is needed, similar to the case of a point of inflection discussed previously. (A point of inflection or a horizontal line segment would be a point of **neutral equilibrium**, as in point D in the previous diagram.)

Another way to look at stable equilibrium is to consider the force. Since F = $-dU/dx$, the force is opposite to the first derivative dU/dx. Note that in this example, dU/dx is proportional to the slope of the tangent line of the curve. (Since y as a function of x represents the curve, dy/dx is the slope of the curve. Since U = mgy, the derivative is dU/dx = mg dy/dx, meaning that dU/dx is proportional to dy/dx, which shows that dU/dx is proportional to the slope of the curve. The proportionality factor is the constant mg.) If the slope of the surface is positive, dU/dx is positive and F is negative. If the slope of the surface is negative, dU/dx is negative and F is positive. Now consider a point of stable equilibrium, which is a local minimum. Such a point is concave up; it looks like a valley. Consider a ball lying at the bottom of this valley. If the ball is displaced slightly to the right of equilibrium, the slope is positive, dU/dx is positive, and F is negative (since F = $-dU/dx$ includes a minus sign); this negative F pushes the ball back to the left (back towards equilibrium). If the ball is displaced slightly to the left of equilibrium, the slope is negative, dU/dx is negative, and F is positive (opposite to dU/dx); this positive F pushes the ball back to the right (back towards equilibrium). Of course, this agrees with common experience. If a ball lies at the bottom of a bowl (like point F in the previous diagram) and you give it a small push, the ball will oscillate back and forth about the point of stable equilibrium at the bottom. In contrast, at a point of unstable equilibrium at the top of a hill (like point B in the previous diagram), if the ball is given a small push, the ball won't return. In this case, if the ball is displaced slightly to the right, dU/dx is negative and F is positive; this F pushes the ball farther to the right (not back towards equilibrium).

Stability is similar in many other cases of equilibrium, including equilibrium thermodynamics or the balancing[77] between radiation pressure and gravitational pull within the sun. Such cases are not as easy to 'see,' and so it is not as easy to intuitively guess what will happen. The mathematics can also be more involved; for example, in equilibrium thermodynamics, there are multiple stability conditions in terms of second derivatives, and the derivatives involve partial derivatives. (In a partial derivative, the function has two or more variables; the derivative is taken with respect to one of the variables while treating the other variables as if they are constant.) But the main idea, that the sign of the second derivative determines whether a critical point is at stable or unstable equilibrium, still applies.

Memory Test (Ch. 11)

Looking at a graph, how can you tell which parts of the graph are concave up or concave down? Which is which?

12 What is optimization?

This is arguably the most practical chapter in this book. When we apply mathematics to the real world, very often it is to minimize or maximize something. Maximizing the profit of a company, making a more fuel-efficient engine, choosing a delivery route that minimizes time, or designing eyeglasses that reduce eye fatigue are examples of the practical value of optimization. Once you express a problem in the form of a function, calculus tells you how to optimize the function. In this chapter, we'll work out one simple example of optimization using numbers, and then discuss a couple of other examples qualitatively.

Suppose that we have the materials to make a fence with a total length of 60 meters. We wish to arrange the fence in the shape of a rectangle, choosing the length and width so that we get the greatest amount of area inside of the fence.

---

[77] Fortunately, this is a point of stable equilibrium, such that when the system gets slightly unbalanced, this is self-correcting, returning to the equilibrium state. Otherwise, life (or even earth) as we know it wouldn't exist.

Let w and x represent the width and length of the fence. The perimeter of the rectangle is P = 2w + 2x and the area of the rectangle is A = xw. It should make sense that the area will be maximized if we use all 60 meters of the fence that is available. This means that the perimeter will be P = 60 meters. Replace P with 60 to get the equation 60 = 2w + 2x. Divide by 2 on both sides to get 30 = w + x. If we subtract x from both sides, we get 30 – x = w. Since 30 – x equals w, this allows us to replace w with 30 – x in the equation for the area: A = x(30 – x). Applying the distributive property, this becomes A = $30x – x^2$. After doing a little algebra, we have a formula for area in terms of x only. Area is a function of x: A(x) = $30x – x^2$. The problem asked us to maximize the area. So, we just need to find the maximum value of the function A(x) = $30x – x^2$.

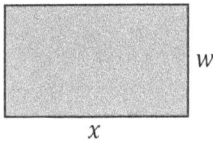

$$\text{perimeter} = x + w + x + w = 2x + 2w$$
$$\text{area} = xw$$

We will use the method from the previous chapter. The first derivative is dA/dx = 30 – 2x. Set the first derivative equal to zero: 30 – 2x = 0. Add 2x to both sides: 30 = 2x. Divide by 2 on both sides: 15 = x. Plug x = 15 into the equation for A(x) to find the corresponding value of area: A(15) = 30(15) – 15^2 = 450 – 225 = 225. The (x, y) coordinates of the critical point are (15, 225). Does it maximize the area? Use the second derivative test. The second derivative is $d^2A/dx^2 = -2$. (We took a derivative of 30 – 2x with respect to x to get this.) Since the second derivative is negative at the critical point (15, 225), this shows that A = 225 square meters is the maximum area. It turns out that the area is maximized when the rectangle becomes a square. When x = 15, w = 30 – x = 30 – 15 = 15, such that the length (x = 15) and width (w = 15) are equal.

If that was more math than you'd like to see, there is good news. In our remaining examples, we will skip most of the math and describe the underlying ideas and process. Here is a recap of the main ideas of the previous example. First, we expressed the area as A = $30x – x^2$. Next, we found the derivative dA/dx = 30 – 2x. We set dA/dx equal to zero to find that x = 15. The derivative dA/dx is zero when the area is an extreme value. We used x = 15 to find that A = 225 square meters. Finally, since the second derivative is negative when x = 15, this shows that A = 225 is a maximum.

Now imagine that a large rectangular box lies on the ground. A rope connects to the center of one side of the box. As indicated below, a worker will pull the rope at an angle. The angle of the rope, θ (the Greek letter theta), indicates how many degrees above the horizontal the rope is pulled. There is friction between the box and the ground, as indicated by the coefficient of friction μ (the Greek letter mu). The box will be pulled so that it travels with constant speed. Which angle θ allows the worker to accomplish this task with the least amount of effort?

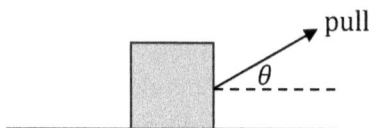

If you're wondering whether the answer is simply zero (meaning that the worker should pull the rope horizontally), it turns out that this is incorrect. Why? If the rope is pulled horizontally, there is more friction between the box and the ground. When θ is positive, the upward component of the worker's pull effectively reduces the amount of friction between the box and the ground. There is sort of a balancing act going on here. A horizontal pull has more horizontal force than a pull at an angle, but a pull at an angle reduces the friction force. Using calculus, we can find exactly which angle minimizes the force with which the worker needs to pull.

The diagram below (called a free-body diagram) illustrates which forces are pulling on the box. There is the force with which the worker pulls on the rope (labeled P), there is the weight of the box pulling downward (labeled mg, since weight is mass times gravitational acceleration), there is the normal force that the ground exerts on the box (labeled N; it's perpendicular to the surface), and there is the force of friction between the ground and the box (labeled f; it's along the ground, opposing the motion).

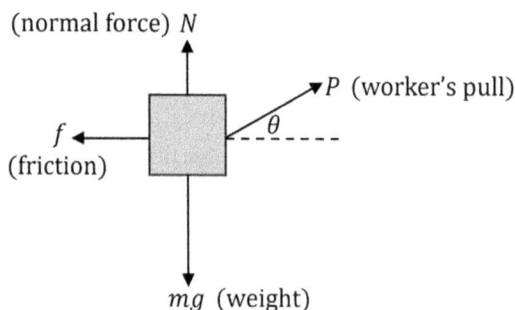

Physics students can apply Newton's second law (which states that the net external force equals mass times acceleration). Using algebra and trigonometry, the student can obtain a formula[78] relating the worker's pull P to the angle θ. Using calculus, the student can set the first derivative of the pull with respect to θ equal to zero to find the value of θ that minimizes P. It turns out that θ equals the inverse tangent of the coefficient of friction: $\theta = \tan^{-1}\mu$.[79] In case you're struggling with the physics concepts and terminology, the main point here is that this is an example of how calculus can help to optimize a problem.

Another example of where optimization is utilized is in thermodynamics, in the context of entropy (which is a measure of statistical disorder) and internal energy (which is the total potential energy plus kinetic energy of its molecules, including rotational, vibrational, and translational kinetic energy; potential energy is stored energy, while kinetic energy is energy of motion). According to the entropy maximum principle of thermodynamics, if the internal energy is held constant, equilibrium is attained when the total entropy is maximized. Given an equation that relates the entropy, internal energy, and other parameters of a system (like volume or the number of molecules), one can apply calculus to find the combination of parameters that maximizes the entropy.

Short Answer (Ch. 12)

Do you remember this concept regarding optimization?

1. Given a function $f(x)$, how can the derivative df/dx help to find the maximum or minimum values of $f(x)$?

---

[78] For those who know physics, the sum of the x-components of the forces gives $P\cos(\theta) - f = ma_x$ and the sum of the y-components of the forces gives $P\sin(\theta) + N - mg = ma_y$ when we apply Newton's second law. Here, $a_y = 0$ because the box doesn't accelerate vertically (assuming that the worker doesn't lift the box right off the ground), which gives $N = mg - P\sin(\theta)$. Multiply by the coefficient of friction (mu) to get the friction force: $f = $ mu times $N = $ mu mg $-$ mu $P\sin(\theta)$. The 'trick' is that $a_x = 0$ because the box is pulled with constant speed. (Acceleration describes how velocity changes, and here velocity doesn't change.) Plug $a_x = 0$ and the equation for f into the x-sum to get $P\cos(\theta) - $ mu mg $+$ mu $P\sin(\theta) = 0$. Isolate P using algebra: $P = $ mu mg $/ [\cos(\theta) + $ mu $\sin(\theta)]$.
[79] Starting with $P = $ mu mg $/ [\cos(\theta) + $ mu $\sin(\theta)]$ from the previous footnote, we can take a calculus shortcut as follows. The numerator (mu mg) is constant. So, P will be maximum when the denominator is minimum. We just need to find the minimum of $\cos(\theta) + $ mu $\sin(\theta)$. Set a derivative with respect to theta equal to zero: $-\sin(\theta) + $ mu $\cos(\theta) = 0$. Add sine to both sides: mu $\cos(\theta) = \sin(\theta)$. Divide by cosine on both sides: mu $= \tan(\theta)$, which means $\theta = $ inverse tangent of mu $= \tan^{-1}\mu$.

13 What is zero divided by zero?

Meet Norma, a fictional child who is really good at arithmetic. (There must be thousands of children like Norma, maybe many more, but we picked Norma because she's entirely fictional.) If you ask Norma what 3 times 8 is, she will promptly answer 24. If you ask her what 30 divided by 5 means, she'll tell you, "Which number times 5 is equal to 30? The answer is 6." Yeah, she's really good at math (like thousands of real children). Let's see if we can stump her with a couple of seemingly simple questions.

Let's ask Norma to find 0 divided by 0. Norma quickly thinks, "Which number times zero is equal to zero?" Then she responds, "The answer is any real number. Any number times zero equals zero. Therefore, zero divided by zero can be anything. It's indeterminate." Wow! Norma is as good as advertised. Many people (not just children) think the answer to 0 divided by 0 is 0, but the answer is actually indeterminate. Let's consider this for a moment, and then we'll try to stump Norma with another question.

When we ask, what is 30 divided by 5, this is equivalent to having 30 apples and organizing them into 5 equal piles. The question is, as Norma said, equivalent to asking which number you can multiply 5 by to make 30. The answer is 6, since 5 times 6 equals 30. If you have 30 apples, you can make 5 equals piles if each pile has 6 apples.

Now compare to the question 0 divided by 0. If we don't have any apples and we don't put them in any piles, it doesn't matter how many apples we put in each pile, the problem is already solved. There are already no apples in no piles without deciding how many to put in each pile. Suppose a poor man decides to put a million dollars in each pile. Since he's poor, he doesn't have a million dollars. But he can say that he put a million dollars in each pile; the only problem is that he has zero piles. The question is equivalent to asking which number you can multiply 0 by to make 0. Well, it turns out that anything you multiply by 0 is equal to 0. Since the answer can be any number, we say that the question is **indeterminate**. We don't know which of the infinite possibilities to give.

It turns out that calculus offers some clues to the question zero divided by zero. It turns out that it depends how the question comes about, as we'll explore later in this chapter. But first, let's ask Norma another simple question, and see if we can't stump her this time. (Maybe we're not really hoping to stump Norma. Maybe we're just trying to challenge her. Maybe her usual homework is just too easy at this stage of her education.)

Let's ask Norma to find 1 divided by 0. Norma swiftly thinks to herself, "Which number times zero equals one?" Then she responds, "No matter what you multiply zero by, it will never be one. Zero times anything is zero. No real number times zero can possibly equal one." Norma is correct again. It doesn't appear that we'll stump her with seemingly simple arithmetic questions after all. (Are there really thousands of real kids like Norma? Maybe, but remember, Norma is fictional.)

Compare with 30 divided by 5. This equates to dividing 30 apples into 5 equal piles. It's equivalent to finding which number you can multiply 5 by to make 30. The answer is 6, since 5 times 6 equals 30. If you have 30 apples, you can make 5 equals piles if each pile has 6 apples.

To find 1 divided by 0, we're trying to put 1 apple in 0 piles. Good luck with that! As Norma said, we want to know which number we can multiply zero by to make one. No matter how many zeros you add together, they'll never add up to one. We say that the question 1 divided by zero is **undefined**. (Okay, so Norma hadn't heard the word 'undefined' before. Her answer is still spot on.) But if we change the question to 1 divided by x and take the limit that x approaches zero, calculus has a different answer than the word 'undefined.'

Now let's see what calculus has to say about dividing by zero. More specifically, we'll consider dividing by x and consider the limit as x approaches zero. As we'll see, the answer to the limit depends on the precise nature of the ratio.

Sometimes, when we take the limit of a ratio, the numerator and denominator each approach finite values. For example, consider the limit as x approaches 2 of the ratio $(x + 3)/(x - 1)$. If you recall what we learned about limits in Chapters 6-7, you should know that the numerator, $x + 3$, approaches $2 + 3 = 5$, while the denominator, $x - 1$, approaches $2 - 1 = 1$, such that the ratio approaches $5/1 = 5$. When the denominator approaches a finite value, the problem is relatively straightforward.

But when the denominator approaches zero, it's not so simple. For example, consider the ratio $\sin(x)$ divided by x. In Chapter 5, we briefly learned about the sine function, and in Chapter 6 we learned that the sine of zero equals zero. In the limit that x approaches zero of the ratio $\sin(x)$ divided by x, we have a problem. The numerator, $\sin(x)$, approaches zero and the denominator, x, also approaches zero. We get the ratio zero divided by zero, which is indeterminate.

Fortunately, when we're exploring limits, there is a way around this indeterminate issue. A special rule, called l'Hôpital's rule, tells us how to evaluate a limit when the numerator and denominator both approach zero. According to **l'Hôpital's rule**, if the numerator and denominator of a ratio each approach zero,[80] you should

---

[80] More generally, l'Hôpital's rule helps whenever the ratio or product of two functions leads to any of the following indeterminate forms: zero over zero, zero times infinity, or infinity divided by infinity.

evaluate the ratio of their derivatives in the same limit.[81] (And if the new numerator and denominator each approach zero even after finding the derivatives, repeat the process until either the numerator or denominator no longer approaches zero.)

Returning to $\sin(x)$ divided by x in the limit that x approaches zero, the derivative of the numerator with respect to x is $\cos(x)$ and the derivative of x with respect to x is one.[82] The ratio of the derivatives is $\cos(x)$ divided by one, which is just $\cos(x)$. Recall from Chapter 6 that $\cos(0) = 1$. Therefore, according to l'Hôpital's rule, the ratio $\sin(x)$ divided by x approaches 1 as x approaches 0. You can see this numerically by entering a small value of x on your calculator. Technically, when a calculus book has $\sin(x)$ divided by x, the book expects you to use radians. A radian is a unit of angle. In particular, π radians = 180 degrees, or 1 radian = $180/\pi$ degrees. So if you want to use $x = 0.01$ radians and your calculator is in degrees mode, you should enter $\sin(0.01$ times $180/\pi) / 0.01$. If you know how to put your calculator in radians mode, you can enter $\sin(0.01)/0.01$. Either way, if you do this correctly, you should get 0.999983333, which is pretty close to one.

Does this mean that $0/0$ equals one? No. It varies from case to case, and it depends on the nature of the functions in the numerator and denominator that led to these limits being zero.

As a second example, consider the ratio $1 - \cos(x)$ divided by x^2 in the limit that x approaches zero. Since $\cos(0) = 1$, the numerator and denominator each approach zero in this limit. Since calculus students know that a derivative of $\cos(x)$ with respect to x equals negative $\sin(x)$, a derivative of the numerator is $\sin(x)$, since the two minus signs cancel.[83] A derivative of the denominator is $2x$. The ratio of the derivatives is $\sin(x)$ divided by $2x$. What happens to this ratio as x approaches zero? The numerator approaches 0 and the denominator approaches $2(0) = 0$. The ratio is still indeterminate. So, we take derivatives again, this time of $\sin(x)$ and $2x$. The derivative of $\sin(x)$ is $\cos(x)$ and the derivative of $2x$ is 2. The new ratio is $\cos(x)$ divided by 2. Since $\cos(0) = 1$, the new ratio approaches $1/2$, which is our final answer. As in the previous example, you can check this on a calculator using 0.01 radians. Enter $(1 - \cos(0.01$ times $180/\pi)) / 0.01^2$

---

[81] This assumes that the derivatives exist at the point in question.

[82] Note that $x = 1x^1$. If you let a = 1 and b = 1, the formula for the derivative gives $(1)(1)x^{1-1}$ = $1x^0 = 1(1) = 1$ because $x^0 = 1$ (see Footnote 32). Alternatively, recall that if y = mx + b, then dy/dx = m. Here, m = 1 and b = 0.

[83] $1 - \cos(x)$ has two terms. The first term, 1, is just a constant, and the derivative of a constant is zero. So we just need the derivative of $-\cos(x)$ with respect to x. Since the derivative of $\cos(x)$ is $-\sin(x)$, the derivative of $-\cos(x)$ is positive $\sin(x)$.

(using two sets of parentheses if your calculator has them; otherwise, enter the calcula-tion for the numerator first and wait until that is done before dividing by 0.01 squared), or if you are using radians mode enter $(1 - \cos (0.01))/0.01^2$. If you do this correctly, you should get 0.49999584, which is pretty close to one-half (since 0.5 = 1/2).

We addressed the issue of 0 divided by 0 (at least, when a ratio approaches this form in a limit) using l'Hôpital's rule. What about the issue of 1 divided by 0? We'll explore this next. It's somewhat simpler, and we saw it in Chapters 6-7.

Consider $f(x) = |1/x|$ (the absolute value of $1/x$) in the limit that x approaches zero. The absolute values ensure that $1/x$ is positive even when x is negative. As x approaches zero, $|1/x|$ grows indefinitely large. We say that the limit as x approaches zero of $f(x)$ equals infinity. (Of course, infinity isn't much like a limit; it's more like the lack of a limit. But we use the word infinity to express the concept that there is no limit to how large $|1/x|$ can get as x gets smaller.) Does this mean that 1/0 equals infinity? Not quite. It's more precise to say that as x gets smaller, $|1/x|$ grows infinite.

Also, note that a limit can approach negative infinity. In fact, if we remove the absolute values, $1/x$ approaches infinity as x approaches zero from above, but $1/x$ approaches negative infinity as x approaches zero from below (recall Chapters 6-7). Without the absolute values, the limit does not exist of $1/x$ as x approaches zero because the limits from below and above don't agree.

Surprise Quiz (Ch. 13)

Can you answer these questions regarding limits? Challenge yourself.

1. Given $f(x) = 1/x$, what is the limit of $f(x)$ as x goes to infinity?

2. If $g(x)$ and $h(x)$ each approach zero as x approaches 1, how can we find the limit of the ratio $g(x)/h(x)$ as x approaches 1?

14 What is an integral?

This is arguably the second most important chapter in this book. Recall that we said Chapter 8, derivatives, was arguably the most important chapter. Many calculus students would probably agree that two of the main things they learned in calculus included derivatives and integrals. The way that calculus is traditionally taught, you need to learn about derivatives before you learn about integrals. Students who have taken calculus know that the process of performing integration can sometimes be challenging; however, the main idea can be understood in reasonably simple terms.

Recall that in Chapter 8 we took a visual approach to derivatives. We found the slopes of tangent lines and then learned that derivatives of functions give you the slopes of tangent lines. In this chapter, we will take a visual approach to integrals. We will explore the area under a curve and see how this relates to a (definite) integral.

Imagine that we have a function $y(x)$, which is a smooth curve without discontinuities, and we wish to find the **area** of the region beneath the curve from x = A to x = B. This problem is illustrated below for a sample curve $y(x)$. The area of interest is shaded. Calculus helps us figure this out without even having to draw the graph, but we'll use the graph as a visual aid to develop what an integral means and how to calculate it (at least, for relatively simple cases).

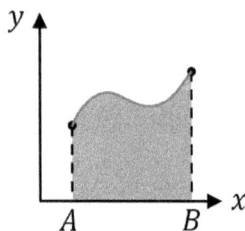

Before we tackle the general problem where $y(x)$ is a curve, let's look at a couple of simple cases where $y(x)$ is a straight line. These cases are so simple we won't even need calculus for them, but we'll be able to use these results to tackle the more general problem.

Suppose that $y(x)$ equals a constant, call it C. The equation for this case is simply $y = C$. This is the equation for a horizontal line. The area below[84] this horizontal line

---

[84] Well, if C is a negative number, it's the area 'above' the curve. In general, we want the area between the curve and the horizontal axis. Calculus students generally say the area 'below' the curve, even though they 'know' that if the curve lies below the horizontal axis, the area is negative and is technically 'above' the curve in that case. We'll discuss this subtle but important point later.

from x = A to x = B is the area of a rectangle with a width of B – A and a height of C. See the diagram below. The area of this rectangle is (B – A)C. For example, if A = 2, B = 8, and C = 5, the width is B – A = 8 – 2 = 6, the height is C = 5, and the area is (6)(5) = 30.

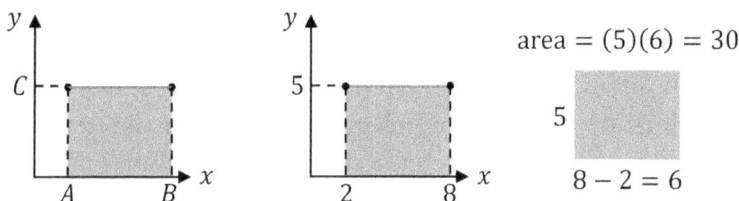

area = (5)(6) = 30

$8 - 2 = 6$

Next, suppose that $y(x)$ is a straight line, such that $y = mx + C$, where m is the slope and C is the y-intercept of the straight line. (Yeah, we normally use b for the y-intercept, but since we're finding the area from x = A to x = B, in order to avoid confusing B with b, we're using C for the y-intercept instead.) This case is illustrated below. In this case, the area below the line is the area of a trapezoid. Oh no, how do you find the area of a trapezoid? Do we need to look up the secret formula? Nope. The trapezoid below is simply a right triangle on top of a rectangle, so all we need to do is add the area of the triangle to the area of the rectangle. See below.

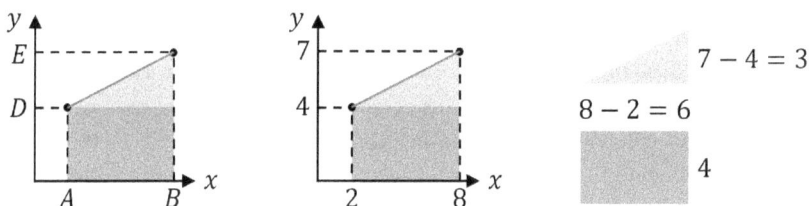

$7 - 4 = 3$

$8 - 2 = 6$

4

We'll do one example with numbers to demonstrate how to find this area (without using the formula for a trapezoid). Suppose that $y = x/2 + 3$ and we wish to find the area from x = 2 to x = 8. This is illustrated above.

- The width of the rectangle and the base of the triangle are 8 – 2 = 6.
- To find the height of the rectangle, plug x = 2 into the formula $y = x/2 + 3$ to get $y = 2/2 + 3 = 1 + 3 = 4$. (This equation is how we found that D = 4 in the picture above.)
- The area of the rectangle is (6)(4) = 24.
- To find the height of the right side of the trapezoid, plug x = 8 into the formula $y = x/2 + 3$ to get $y = 8/2 + 3 = 4 + 3 = 7$. (This equation is how we found that E = 7 in the picture above.)

• To find the height of the triangle, subtract the height of the rectangle from the height of the trapezoid: 7 − 4 = 3.

• Recall from the first bullet point that the base of the triangle (and rectangle) is 6 and recall from the previous bullet point that the height of the triangle (NOT the rectangle and NOT the trapezoid) is 3. The area of the triangle is one-half its base times its height: (1/2)(6)(3) = 9.

• Add the area of the triangle and rectangle together to find the area of the trapezoid: 24 + 9 = 33. (If you look up the formula for the area of a trapezoid, that would work, too, and save a few steps. We said it wasn't necessary to know that formula, but we never claimed that it wouldn't be useful.)

Now we're ready to consider the more general case where $y(x)$ is a smooth curve rather than a straight line. Any ideas? One approach is to divide the curve into a large number of skinny rectangles, as illustrated below.[85] When we have a small number of rectangles (like the left graph below), the area of the rectangles is noticeably different from the area under the curve; the gaps between the curve and the rectangles are pronounced. But if we make a larger number of skinnier rectangles, the area of the rectangles[86] (all added together, of course) becomes a better match for the area under

---

[85] How should we draw the rectangles? We could shorten the rectangles so that the top of each rectangle lines up with the lowest part of the curve within the width of each rectangle; such a rectangle is said to be inscribed. We could line up the top/center of each rectangle with the corresponding point on the curve. We could match the top/right of each rectangle with the top of each curve. There are a lot of ways to do it, and if you consult a calculus textbook, you might see it drawn a bit differently. Since for an integral we will eventually take the limit that the width of the rectangle approaches zero, it won't really matter. For an infinitesimally thin rectangle, there is no difference between the top center of the rectangle and the top left or top right of the rectangle; it's not thick enough to notice any difference. But where it does matter is in numerical approximations to integrals. In that case, this is a very big issue. For that, it's better to use a trapezoid than a rectangle, so that the shape matches the curve better. With numerical approximations, a finite number of rectangles (or trapezoids) is used which each have finite width. For an integral, there are an infinite number of infinitesimally thin rectangles, for which shape doesn't matter. For numerical approximations, shape very much does matter.

[86] Wouldn't it be better to use trapezoids instead of rectangles? Yes, if the number of trapezoids is finite, it would be. When calculus students learn how to approximate definite integrals numerically, they use trapezoids instead of rectangles. (Then they learn Simpson's rule, which uses parabolas instead of trapezoids.) But in calculus, an integral is performed by taking the limit that the width of the rectangles approaches zero. In this limit, the rectangle or trapezoid

the curve. This is the main idea behind integral calculus.

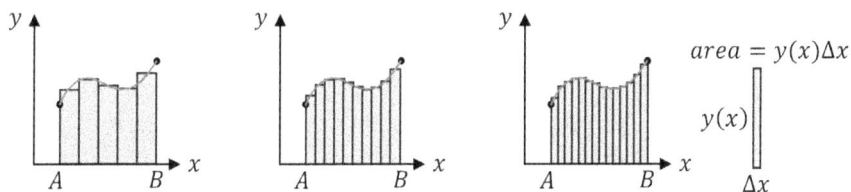

First, suppose that we divide the region below[87] the curve up into N rectangles, where each rectangle has the same width called Delta x. (Calculus students write this as Δx, where Δ is the uppercase Greek letter Delta. We'll write out the word Delta so that you don't have to memorize the name for the Greek letter Δ.) The area of each rectangle is $y(x)$ times Delta x, since $y(x)$ is the height of the rectangle for a particular[88] value of x. The total area of the rectangles is the sum of $y(x)$ times Delta x for each rectangle. When N is finite, the width of each rectangle Delta x is finite, and the total area of the rectangles differs from the actual area below the curve. But in the limit that N grows infinite, the width of each rectangle becomes infinitesimal. The width of each rectangle is a differential element (a term we learned in Chapter 8 on derivatives; a differential element has infinitesimal size). When the width is finite, we call it Delta x, but when it is infinitesimal, we call it dx, since it is a differential element. Taking the limit as Delta x approaches zero, the sum has an infinite number of terms $y(x)$ times dx added together and the value of this sum is equal to the area under the curve. We call this particular infinite sum (of infinitesimally thin rectangles) a **Riemann sum**, and it serves as the basis for finding the area under a curve.

Obviously, you can't add an infinite number of terms together. Fortunately, calculus tells us how to determine the infinite sum of infinitesimal rectangles without having to add them all together one at a time. We'll now turn our attention to how to do this. In calculus, it's called an integral. In particular, the **integral** of $y(x)$ dx

---

is so skinny that it doesn't make a difference. We actually get an exact answer using infinitesimal rectangles, so we don't need to bother with trapezoids (or parabolas) in this limit.

[87] That's if the curve lies above the horizontal axis. If a curve lies below the horizontal axis, then the area is above the curve, as we will see later in the chapter.

[88] The value of x is somewhere within the width of the rectangle. When the rectangle has finite width, it's important to decide exactly where to measure x (for example, to the left side of the rectangle, to the center, or the textbook choice of the point in the rectangle where the curve is lowest), but since we'll be taking the limit that the width of the rectangle approaches zero, in that limit, there won't actually be a choice.

from the initial value x = A to the final value x = B represents the area between the curve $y(x)$ and the horizontal axis. This integral is equivalent to the Riemann sum that we just discussed. That is, the integral of $y(x)$ dx from x = A to x = B gives the same value as adding up the areas of an infinite number of infinitesimally thin rectangles spanning from A to B. It equals the limit as Delta x approaches zero of the sum of $y(x)$ times dx from A to B. And it equals the area between the curve $y(x)$ and the x-axis. All of these are equivalent.

The notation for an **integral** is shown below. The large symbol at the left is called the integration symbol (or the integral sign). The function $y(x)$ is called the integrand; it's the function being integrated over. The differential element dx tells us which variable we're integrating over; in this case, we're integrating over the variable x (or you could say that we're integrating with respect to x). The x = A underneath the integration symbol tells us that the integral begins at x = A, and the B above the integration symbol tells us that the integral ends at x = B. The process of evaluating the integral to see what it equals is called **integration**. The endpoints x = A and x = B are called the limits of integration. Calculus teaches a variety of techniques of integration for evaluating different kinds of integrals, but presently we'll focus on the main general idea of how to find the integral.

$$\int_{x=A}^{B} y(x)\, dx$$

The integral above, which has lower and upper limits, is called a **definite integral**. The value of a definite integral is equal to a numerical value representing the area between $y(x)$ and the x-axis. The integral below, which doesn't have integration limits, is called an **indefinite integral**. The answer to an indefinite integral is a function (and as we'll discuss in Chapter 15, it includes a constant of integration).

$$\int y(x)\, dx$$

It's important to understand that the variable of integration doesn't really matter. We could call it x, t, u, or any letter and the value of the integral would be the same. The three definite integrals below are the same. The function is the same each time, except for being written in terms of a different symbol. For example, if $y(x)$ = x^2, then $y(t)$ = t^2 and $y(u)$ = u^2. Whether we name the variable x, t, u, or any other letter, it doesn't matter. The value of the integral and the area under the curve are the same either way. All that really matters is that we don't choose a letter that would be easily confused (or the same as) some other letter that we're using for something else in the same problem.

$$\int_{x=A}^{B} y(x)\, dx = \int_{t=A}^{B} y(t)\, dt = \int_{u=A}^{B} y(u)\, du$$

How do we find the value of an integral without adding up an infinite number of areas of infinitesimally thin rectangles? It turns out that the answer is related to derivatives. In Chapter 8, we learned that derivatives relate to the slope of a tangent line, and in this chapter we're learning that integrals relate to the area under a curve. These two different concepts and approaches, derivatives and integrals, turn out to be related, even though at first they may seem quite different. They are related through the first fundamental theorem of calculus. We'll discuss (but not prove) this theorem now along with the important result that it provides.

Suppose that we're integrating the function y(t) dt from t = A to t = x. This should seem a bit funny at first. Here, the variable of integration is t and the upper limit includes a different variable x. There is a reason for this; since the upper limit of integration is the variable x, our answer for the definite integral in this case will be a function of x. The area under the curve depends on the upper limit x. (Eventually, we will learn how to find the integral when the limits are constants, but we need to do this presently to explain what the first fundamental theorem of calculus is and how it will help us perform integrals.) Let f(x) be the function that we obtain as a result of the integration. The integral and its result are shown in the equation below.

$$f(x) = \int_{t=A}^{x} y(t)\, dt$$

According to the **first fundamental theorem of calculus**, if y(t) is continuous over the interval (starting at t = A), then the result of the integration, f(x), is differentiable at any value of x over the interval and it turns out that a derivative of f(x) with respect to x is equal to y(x). This is an important and powerful result, so we'll take a moment to highlight the main points here.

• y is the function we're integrating over. (It's the integrand.) Whether we write it as y(t) or y(x) doesn't really matter. It has the same functional form regardless. For example, if y(t) = t^2, then y(x) – x^2. They are the same except for the letter used for the argument. If you find y(5), regardless of whether you use y(t) or y(x), you'll get the same value: y(5) = 5^2 = 25.

- f(x) is the result of the integration. The function f(x) is called the **antiderivative**. The reason for this name will soon become clear.
- The function y(x) in the integrand and the function f(x) that is the result of the integration are related. In particular, y(x) = df/dx.
- The function y(x) is a derivative of f(x) with respect to x.
- The function f(x) is the **antiderivative** of y(x). It's basically the opposite (or the inverse[89] operation) of a derivative.
- The key point is that we'll be able to perform integrals by finding anti-derivatives. Once we learn exactly what an antiderivative means and how to find it, we'll have a strategy for how to perform an integral.

If y(x) is a derivative of f(x) with respect to x, that is y = df/dx, then f(x) is the **antiderivative** of y(x). Given a function y(x), to find its antiderivative f(x), we must think 'backwards' compared to how we think about derivatives. In particular, given y(x), to find f(x), we ask, "Which function could you take a derivative of and obtain y(x) as a result." We'll illustrate this with a couple of examples.

Suppose that y(x) = 2x and we wish to find its antiderivative, which we will call f(x). We're asking, "Which function f(x) could you take a derivative of with respect to x and obtain 2x as the answer?" One such function is f(x) = x^2. If you take a derivative of f(x) = x^2 with respect to x using the technique from Chapter 8, you'll get[90] df/dx = 2x. The function f(x) = x^2 is the antiderivative of y(x) = 2x. Put another way, y(x) = 2x is the derivative of f(x) = x^2 with respect to x. The two ways of thinking are related. If we have y(x) = 2x and wish to find its antiderivative, the answer is f(x) = x^2. If instead we have f(x) = x^2 and wish to find its derivative, the answer is y(x) = 2x.

---

[89] It's not the multiplicative inverse. There are different kinds of inverses in algebra and calculus. Students are most familiar with the term 'inverse' that they learn in algebra, which means to find the reciprocal, but the term inverse is used more generally than that. Here, inverse doesn't mean reciprocal. Recall the trig functions that we learned about in Chapter 5. The inverse sine function, $\sin^{-1}(x)$, for example, doesn't mean to find the reciprocal (which would be cosecant). Rather, $\sin^{-1}(x)$ means, "Which angle could you take the sine of an obtain the fraction x as a result?" If you act with a function and then act with its inverse, the two operations cancel out. In multiplication, the inverse is a reciprocal because x times x^{-1} equals one; the two factors cancel out. In trig, the $\sin(\sin^{-1}(x)) = x$. The sine of the inverse sine cancels out. In calculus, the derivative of the antiderivative leaves the function unchanged. If you first find the antiderivative of a function and then find the derivative of that, you get the original function back. We'll see an example of this shortly.
[90] Compare x^2 with ax^b to see that a = 1 and b = 2. Using the formula from Chapter 8, abx^{b-1} = (1)(2) x^{2-1} = $2x^1$ = 2x.

But $f(x) = x^2$ isn't the only possible answer. We could just as well say that $f(x) = x^2 + 3$ is the antiderivative of $y(x) = 2x$. Why? If we take a derivative of $f(x) = x^2 + 3$ with respect to x, we get the same answer, $y(x) = 2x$, because the derivative of a constant is zero. In fact, we can add any constant we want to $f(x) = x^2$ and the new function, $f(x) = x^2 + C$ (where C is a constant), will be the antiderivative of $y(x) = 2x$. This constant is called a **constant of integration**. We'll discuss the constant of integration in Chapter 15, and focus on the main form of the antiderivative for now.

As a second example, suppose that $u(x) = x^3$ and its antiderivative is $g(x)$. We're asking, "Which function $g(x)$ could you take a derivative of with respect to x and obtain x^3 as the answer?" One such function is $g(x) = x^4/4$. (How are we figuring out what the answer is? Good question. We'll address that soon.) Again, we can add a constant of integration to $g(x)$, but we'll save that for Chapter 15. If you take a derivative of $g(x) = x^4/4$ with respect to x, using the technique from Chapter 8, a = 1/4 and b = 4, which gives us $u(x) = dg/dx = abx^{b-1} = (1/4)(4)\, x^{4-1} = 1x^3 = x^3$.

It's easy to find the antiderivative of a polynomial. We can work out a formula for it by understanding how to find the derivative of a polynomial. So first we'll quickly review the formula for the derivative of a polynomial term, and use it to find a formula for the antiderivative of a polynomial term.

Recall from Chapter 8 that for a polynomial term of the general form ax^b, the derivative has the form abx^{b-1}. That is, when we find the derivative of a term of a polynomial, the exponent comes out to multiply the coefficient and then we reduce the exponent by one. For example, to find a derivative of $6x^8$ with respect to x, the 8 comes out to multiply the 6 and the exponent 8 is reduced to 7. Thus, the derivative of $6x^8$ with respect to x is $(6)(8)\, x^7 = 48x^7$. Using a = 6 and b = 8, this agrees with the formula abx^{b-1}.

The formula for the antiderivative of a term of a polynomial is very similar. This time, the exponent is raised by one (instead of reduced by one), and the coefficient is divided by the new exponent (instead of being multiplied by the original exponent). Specifically, to find the **antiderivative** of a polynomial term of the general form ax^b, use the formula $\dfrac{ax^{b+1}}{b+1}$, with one important exception. If b = -1, we'll learn a different formula later in the chapter. (You may also add a constant of integration to this.) For example, to find the antiderivative of $48x^7$, identify a = 48 and b = 7. The formula gives us $\dfrac{48x^{7+1}}{7+1} = \dfrac{48x^8}{8} = 6x^8$ (since 48 divided by 8 is equal to 6).

Observe that the two previous paragraphs are related. A derivative of $6x^8$ with respect to x is $48x^7$. The antiderivative of $48x'$ is $6x^0$ (plus a constant of integration). This is the sense in which an antiderivative is 'backwards' (or the inverse, but not the multiplicative kind) compared to a derivative.

The way to evaluate an **integral** is to find the **antiderivative** of the function in the integrand. This is basically what the first fundamental theorem of calculus lets us do: If the function y(t) in the integrand below

$$f(x) = \int_{t=A}^{x} y(t)\, dt$$

is continuous over the interval (starting at t = A), then the result of the integration f(x) is differentiable at any value of x over the interval and it turns out that y(x) and f(x) are related by the derivative y(x) = df/dx. The equation y(x) = df/dx tells us that f(x) is the antiderivative of y(x). Finding f(x) is the key to performing the integral.

Now we will provide the method for finding a **definite integral**, like the one below. Unlike an indefinite integral, a definite integral has limits below and above the integration symbol; the example below begins at x = 1 and ends at x = 4. The second fundamental theorem of calculus tells us how to perform a definite integral. According to the **second fundamental theorem of calculus**, if y(x) is continuous over the interval (from x = A to x = B) and if f(x) is the antiderivative of y(x) along the interval, then the result of the definite integral is equal to f(B) – f(A). The second fundamental theorem of calculus allows us to perform a definite integral in three steps:

• Given the integrand y(x), first find its antiderivative f(x). This step is the same as finding the indefinite integral of y(x). (However, unlike an indefinite integral, we won't need the constant of integration. Why not? Because the constant of integration would cancel out in the last step.)

• Evaluate the antiderivative, f(x), at each integration limit. That is, find f(B) and f(A).

• Subtract. The final answer is f(B) – f(A). The final answer is one number. (This number equals the area between the curve and the horizontal axis along this interval.)

$$\int_{x=1}^{4} y(x)\, dx$$

Let's quickly compare the strategies for indefinite and definite integrals. The answer to an **indefinite integral** like the one below is the **antiderivative** of the function in the integrand (including a constant of integration). An indefinite integral doesn't include integration limits. The final answer for an indefinite integral is a function.

$$\int y(x)\, dx$$

In contrast, a definite integral includes limits of integration below and above the integration symbol; the example below begins at $x = 2$ and ends at $x = 6$. The answer to a **definite integral** like the one below is found by first finding the antiderivative of the function in the integrand, evaluating the antiderivative at each endpoint, and subtractting. The final answer for a definite integral is just a numerical value (not a function of x). In the remainder of the chapter, we'll work out examples of definite and indefinite integrals with a variety of specific functions in place of the more general function $y(x)$.

$$\int_{x=2}^{6} y(x)\, dx$$

As an example, consider the indefinite integral below. The integrand is $y(x) = 12x^3$ and x is the variable of integration. To perform the indefinite integral, we just need to find the antiderivative of the function $y(x) = 12x^3$. Compare $12x^3$ with ax^b to identify $a = 12$ and $b = 3$. Use the formula for the antiderivative (not to be confused with the formula for the derivative from Chapter 8): $\dfrac{ax^{b+1}}{b+1} = \dfrac{12x^{3+1}}{3+1} = \dfrac{12x^4}{4} = 3x^4$. (In the last step, we divided 12 by 4 to make 3.) Since the answer to an antiderivative or indefinite integral in general may include a constant of integration (the subject of the next chapter), our final answer is $3x^4 + C$. As expected, for an indefinite integral, the answer is a function of the variable of integration (in this case, that's x).

$$\int 12x^3\, dx$$

It's not always easy to perform an integral, but it's always straightforward to check the answer. Why? Finding a derivative is generally easier than finding an antiderivative. With simple polynomial terms, each is equally easy, but with more general expressions, derivatives tend to be more straightforward than antiderivatives. In our current example, if you take a derivative of $3x^4 + C$ with respect to x, you get $(3)(4)x^{4-1} + 0 = 12x^3$. Since our integrand was $12x^3$, our answer checks out.

Now consider the definite integral below. It's the same integral as our previous example, except that this integral is a definite integral because it includes limits. The first step to performing a definite integral is to find the antiderivative. Since the integrand $12x^3$ is the same as before, the antiderivative will still be $3x^4$. (For a definite integral, there is no reason to include the constant of integration. Why? Because we will be subtracting two terms. Any constant of integration would simply cancel out in the subtraction.) The next step is to evaluate the antiderivative at each limit. We'll use the upper limit first. Plug $x = 2$ into $3x^4$ to get $3(2)^4 = 3(16) = 48$. Now use the lower limit. Plug $x = 1$ into $3x^4$ to get $3(1)^4 = 3(1) = 3$. To get the final answer, simply subtract 3 from 48. We get $48 - 3 = 45$. The final answer is 45. For a definite integral, the answer is just a number. The area under the curve $12x^3$ from $x = 1$ to $x = 2$ is exactly equal to 45. We found this area by finding the antiderivative of the function $12x^3$. This is one of the main principles of calculus.

$$\int_{x=1}^{2} 12x^3 \, dx$$

Let's discuss a point that we alluded to earlier, but which we haven't yet fully addressed. Calculus students often remember that a definite integral represents the area 'under' a curve, but this isn't quite precise. It's more precise to say that a definite integral represents the area between a curve and the horizontal axis. When the curve is above the horizontal axis, then it is indeed the area 'under' the curve. But when the curve is below the horizontal axis, then it is the area 'above' the curve. Either way, it's the area between the curve and the horizontal axis. With a caveat. If the curve lies below the horizontal axis, the 'area' is negative.

For example, consider the diagram below. From $x = 0$ to $x = 4$, the function $y(x)$ lies below the x-axis; in this region, the integral is negative. From $x = 4$ to $x = 10$, the function $y(x)$ lies above the x-axis; in this region, the integral is positive.

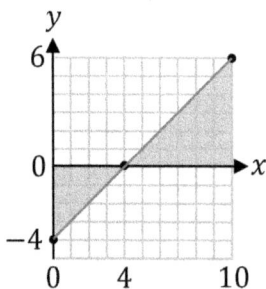

The function in this example, $y(x) = x - 4$, is a straight line. We'll find the definite integral of $y(x)$ dx in each region to see that one area is positive while the other area is negative, and then we'll see that we can actually find the area from $x = 0$ to $x = 10$ in a single step (without having to find two separate areas). Also, since each area happens to be a right-triangle in this example, you could use the formula that the area of a triangle is one-half base times height, in case you'd like to check that the calculus we do below agrees with what you already know from geometry.[91] (Why are we even bothering with calculus instead of using geometry? Because if the function is a parabola, quarter ellipse, or some other function that isn't a straight line, you wouldn't be able to use the formula one-half base times height. Calculus offers a general method for finding the area below a curve.) The antiderivative of $y(x) = x - 4$ is equal to $x^2/2 - 4x$ (plus a constant of integration).[92]

- To find the integral from $x = 0$ to $x = 4$, first plug $x = 4$ into the antiderivative $x^2/2 - 4x$ to get $4^2/2 - 4(4) = 16/2 - 16 = 8 - 16 = -8$. Next, plug $x = 0$ into the antiderivative to get $0^2/2 - 4(0) = 0 - 0 = 0$. Now subtract these two values (starting with the upper limit): $-8 - 0 = -8$. The definite integral is indeed negative in this region. Since $y(x)$ is a straight line, we can easily check the area. The area below $y(x)$ in this region is a triangle with a base of 4 and a height of 4. Its area is $(1/2)(4)(4) = (1/2)(16) = 8$. Thus, we see that when $y(x)$ lies below the x-axis, our method of integration gives us an answer equal to the negative of the area above $y(x)$.

- To find the integral from $x = 4$ to $x = 10$, first plug $x = 10$ into the antiderivative $x^2/2 - 4x$ to get $10^2/2 - 4(10) = 100/2 - 40 = 50 - 40 = 10$. Next plug in $x = 4$. We did that in the previous bullet point and found that the antiderivative was -8 when $x = 4$. Subtract these two values (starting with the upper limit): $10 - (-8) = 10 + 8 = 18$. The definite integral is positive in this region. Let's check the area using the fact that $y(x)$ is a straight line. The area above $y(x)$ in this region is a triangle with a base of 6 (look at the graph above) and a height of 6. Its area is $(1/2)(6)(6) = (1/2)(36) = 18$. When $y(x)$ lies above the x-axis, our method of integration gives us an answer equal to the area below $y(x)$.

---

[91] The first triangle has a base of 4 (since it extends from $x = 0$ to $x = 4$) and a height of 4 (since the y-intercept is -4), so the first triangle has an area of $(1/2)(4)(4) = 8$. It's really -8, since this area is below the x-axis. The second triangle has a base of 6. To find its height, plug in $x = 10$ to get $y(10) = 10 - 4 = 6$. The second triangle has a height of 6. The area of the second triangle is $(1/2)(6)(6) = 18$. The total area from $x = 0$ to $x = 10$ is the sum of these areas, which is $-8 + 18 = 10$. If you compare these results from the geometry of right triangles to the calculus that we do in the text, you'll see that they agree.

[92] This is easy to check. First, as we'll learn later in this chapter, just like with derivatives, when there are two or more terms, we can integrate each term separately. A derivative of x^2 with respect to x is $2x$, so a derivative of $x^2/2$ with respect to x is simply x. A derivative of $4x$ with respect to x is 4. Combining these ideas together, a derivative of $x^2/2 - 4x$ with respect to x is $x - 4$.

• The total area from $x = 0$ to $x = 10$ is the sum of our two previous answers: $-8 + 18 = 10$. But if you want the definite integral from $x = 0$ to 10, **there is absolutely no reason to divide the integral into two pieces** as we've done. A beautiful feature of calculus is that you can find the definite integral from $x = 0$ to $x = 10$ without dividing the integral into separate pieces for the positive and negative areas. We'll demonstrate this now. To find the integral from $x = 0$ to $x = 10$, we'd do the same as before. First plug $x = 10$ into the antiderivative, which gave us a value of 10 in the previous bullet point. Next plug $x = 0$ into the antiderivative, which gave us a value of 0 in the first bullet point. Now subtract: $10 - 0 = 0$. See, we got the same answer for the definite integral from $x = 0$ to $x = 10$ in a single step this way.

Now let's discuss the exception to the antiderivative of a function of the form ax^b. The rule that we learned, that the antiderivative is $\frac{ax^{b+1}}{b+1}$, only applies if b doesn't equal -1. In the case that $b = -1$, it's different. First of all, if you tried plugging $b = -1$ into the formula, you'd run into a problem of dividing by zero, since $b + 1$ is equal to $-1 + 1 = 0$ in that case. But that's aside from the main point. To better understand the case where $b = -1$, recall the rule from algebra that $x^{-1} = 1/x$. In the case $b = -1$, the function in the integrand has the form $ax^b = ax^{-1} = a/x$. This antiderivative problem asks, "Which function can you take a derivative and obtain a/x as a result?"

Can you recall a function from Chapter 8 (regarding derivatives) where we took a derivative and the answer was $1/x$? There was one: the natural logarithm. Recall from Chapter 8 that if $f(x) = \ln(x)$, then $df/dx = 1/x$. This means that the antiderivative of $1/x$ is equal to $\ln(x)$. Technically, it's $\ln|x|$ using absolute values, but that's more technical than we intend to get in this book. (Of course, you may also add on a constant of integration.) So if an integrand has the form ax^b, the indefinite integral (and thus the antiderivative) is equal to $\frac{ax^{b+1}}{b+1}$ plus a constant of integration if b isn't equal to -1 and is equal to a times $\ln|x|$ plus a constant of integration if b does equal -1. This is the complete rule for integrands of the form ax^b.

$$\int ax^b \, dx = \begin{cases} \dfrac{ax^{b+1}}{b+1} + c & \text{if } b \neq -1 \\ a \ln|x| + c & \text{if } b = -1 \end{cases}$$

Do you recall the special thing about exponentials from when we learned about derivatives? In case you forgot, the simple exponential e^x is its own derivative. If $f(x)$

$= e^x$, then $df/dx = e^x$. Since e^x is its own derivative, it's also its own antiderivative (apart from a possible constant of integration). If the exponent includes a constant as in e^{kx}, where k is a constant, then the k comes out in the derivative. That is, if $g(x) = e^{kx}$, then[93] $dg/dx = ke^{kx}$. (This is a consequence of the chain rule that we learned about in Chapter 8.) In the case of an antiderivative, $1/k$ comes out. That is, if $g(x) = e^{kx}$, then the antiderivative of $g(x)$ is $\frac{e^{kx}}{k}$. (This way, if you take a derivative of the antiderivative, you get the same function back.)

$$\int e^x \, dx = e^x + c \quad , \quad \int e^{kx} \, dx = \frac{e^{kx}}{k} + c$$

The two simplest trig functions to integrate are sine and cosine. In Chapter 8, we learned that the derivative of $\sin(x)$ is $\cos(x)$. Therefore, the antiderivative of $\cos(x)$ is $\sin(x)$. In Chapter 8, we also learned that the derivative of $\cos(x)$ is $-\sin(x)$. Therefore, the antiderivative of $\sin(x)$ is $-\cos(x)$.

$$\int \cos x \, dx = \sin x + c \quad , \quad \int \sin x \, dx = -\cos x + c$$

When we learned about derivatives, we learned that there were two equivalent ways of thinking about derivatives. One way (which we learned in Chapter 8) is that the derivative of a function represents the slope of a tangent line. Another way (which we learned in Chapter 9 and which was also discussed in Chapter 3) is that a derivative is an instantaneous rate of change. For example, velocity is the instantaneous rate at which position changes in time, and velocity is a derivative of position with respect to time. In this case, we took the limit as the time interval approached zero of the change in position divided by the time interval. If you plot position as a function of time, this same concept can be visually interpreted as the slope of a tangent line. These two ways of thinking about a derivative are equivalent in that each leads to the same limit as time approaches zero.

We can similarly think about integrals two different ways. One way, which we've already explored, is the area between a curve and a horizontal axis. Another way relates to an instantaneous rate of change. We'll explore this second way presently.

A second way of thinking about an integral can be understood by considering that velocity (which is a combination of speed and direction) is a derivative of position

---

[93] Wait! Why did we switch the notation from $f(x)$ to $g(x)$? We changed the name because the function is different. Compare $f(x) = e^x$ with $g(x) = e^{kx}$. These are two different functions. One has a k in the exponent.

with respect to time: For straight-line motion along the x-axis, we express this as $v_x = dx/dt$. We're using v_x to represent velocity.[94] If you know the position as a function of time, you could find velocity by taking a derivative of x with respect to t. Suppose instead that we know velocity as a function of time, and wish to find position. How would you do that? The answer is to find an antiderivative. If v_x is a derivative of x with respect to t, then x is an antiderivative of v_x. The way to find v_x from x is to perform the integral below.

$$x = \int v_x \, dt$$

With the indefinite integral, there is a constant of integration. If instead we perform a definite integral, what we get is the **net displacement**, which is a straight line from the initial position to the final position. **Net displacement** represents the change in the object's position.

$$\text{net displacement} = \int_{t=t_i}^{t_f} v_x \, dt$$

Velocity is a derivative of position with respect to time, and acceleration is a derivative of velocity with respect to time: $a_x = dv_x/dt$. If you want to find velocity from acceleration, you need to integrate. If you express the integral as a definite integral, it tells you the change in velocity: the final velocity v_f minus the initial velocity v_i.

$$v_f - v_i = \int_{t=t_i}^{t_f} a_x \, dt$$

---

[94] In three-dimensional motion, velocity (which is a vector; it includes both speed and direction) has components along the x-, y-, and z-axes. We label these components v_x, v_y, and v_z. In straight-line motion along the + or − x-axis, there is just one component, v_x. Note that v_x may be positive or negative; it would be negative for an object moving in the negative x-direction.

Try It Yourself (Ch. 14)

There are two straightforward exercises below designed to help you feel like you're doing some real calculus. Following the exercises, you can find the full solution to each exercise with explanations. But you can probably solve the exercises by yourself just by studying the example. You can do it.

Example. Perform the definite integral below.

$$\int_{x=2}^{3} 10x^4 \, dx$$

Compare ax^b with $10x^4$ to identify a = 10 and b = 4. Plug these into the formula $\frac{ax^{b+1}}{b+1}$ to get $\frac{10x^{4+1}}{4+1} = \frac{10x^5}{5} = 2x^5$. Plug the upper limit x = 3 into $2x^5$ to get $2(3)^5 = 2(243)$ = 486. Plug the lower limit x = 2 into $2x^5$ to get $2(2)^5 = 2(32) = 64$. Subtract 64 from 486. Our final answer is $486 - 64 = 422$.

Directions. Perform the definite integrals below.

❶ $$\int_{x=1}^{2} 6x^5 \, dx$$

❷ $$\int_{x=4}^{5} (15x^2 - 4) \, dx$$

Note: These integrals are relatively straightforward compared to many challenging integrals that calculus students learn how to solve. The exercises above offer a small taste of what calculus is like.

Solutions to the Ch. 14 Exercises

1. Compare ax^b with $6x^5$ to identify $a = 6$ and $b = 5$. Plug these into the formula $\frac{ax^{b+1}}{b+1}$ to get $\frac{6x^{5+1}}{5+1} = \frac{6x^6}{6} = x^6$. Plug the upper limit $x = 2$ into x^6 to get $2^6 = 64$. Plug the lower limit $x = 1$ into x^6 to get $1^6 = 1$. Subtract 1 from 64. Our final answer is $64 - 1 = 63$.

$$\text{❶} \qquad \int_{x=1}^{2} 6x^5\, dx = [x^6]_{x=1}^{2} = 2^6 - 1^6 = 64 - 1 = 63$$

2. Treat each term separately. First find the antiderivative of $15x^2$. Compare ax^b with $15x^2$ to identify $a = 15$ and $b = 2$. Plug these into the formula $\frac{ax^{b+1}}{b+1}$ to get $\frac{15x^{2+1}}{2+1} = \frac{15x^3}{3} = 5x^3$. Now find the antiderivative of 4. The antiderivative of the constant 4 is simply $4x$. (Why? Because a derivative of $4x$ with respect to x equals 4. If you want to use the formula, use $a = 4$ and $b = 0$, since $x^0 = 1$.) The antiderivative of $15x^2 - 4$ is equal to $5x^3 - 4x$. (Since this is a definite integral, we don't need to include a constant of integration, as mentioned in the text of this chapter.) Plug the upper limit $x = 5$ into $5x^3 - 4x$ to get $5(5)^3 - 4(5) = 5(125) - 20 = 625 - 20 = 605$. Plug the lower limit $x = 4$ into $5x^3 - 4x$ to get $5(4)^3 - 4(4) = 5(64) - 16 = 320 - 16 = 304$. Subtract 304 from 605. Our final answer is $605 - 304 = 301$.

$$\text{❷} \qquad \int_{x=4}^{5} (15x^2 - 4)\, dx = [5x^3 - 4x]_{x=4}^{5}$$

$$= 5(5)^3 - 4(5) - [5(4)^3 - 4(4)]$$

$$= 5(125) - 20 - [5(64) - 16]$$

$$= 625 - 20 - (320 - 16) = 605 - 304 = 301$$

15 What is a constant of integration?

As we learned in the previous chapter, an indefinite integral is performed by finding an antiderivative and includes a constant of integration. Let's quickly recall why there is a constant of integration, and then we'll discuss what it means and how it is used.

It's easiest to understand the constant of integration with a specific example in mind. Consider the derivative of $4x^2$ with respect to x. To find the derivative, identify a = 4 and b = 2, and the use the formula (not to be confused with the formula for an antiderivative) $abx^{b-1} = (4)(2)x^{2-1} = 8x^1 = 8x$. Now consider the derivative of $4x^2 + 7$ with respect to x. We get the same answer for the derivative, $8x$, because the second term is a constant; a derivative of 7 with respect to x is zero. The derivatives of $4x^2$, $4x^2 + 7$, $4x^2 - 3$, $4x^2 + 100$, and $4x^2 + c$ (where c is a constant) are all equal to $8x$.

Now consider this example in reverse. Suppose that we are given $8x$ and wish to find its antiderivative. There isn't just one answer for the antiderivative. It could be $4x^2$, $4x^2 + 7$, $4x^2 - 3$, $4x^2 + 100$, or $4x^2$ + any other constant. Ultimately, the reason that an indefinite integral (or antiderivative) includes a constant of integration is that the derivative of a constant equals zero. When finding an indefinite integral, we're asking, "Which function could we take a derivative of and obtain the integrand as a result?" (In the previous example, $8x$ is the integrand.) Whatever that function is, adding a constant to it will still give the same integrand back if we then take a derivative (because the derivative of a constant is zero). The main idea is that the indefinite integral could be $4x^2$, $4x^2 + 7$, $4x^2 - 3$, $4x^2 + 100$, or $4x^2$ + any other constant. Therefore, to cover the general case, we write the final answer to the indefinite integral as $4x^2$ + c.

$$\int 8x \, dx = 4x^2 + c$$

The indefinite integral basically has an indefinite answer. We don't know what the constant of integration, c, is (without more information). In contrast, a definite integral has a definite answer; it's a single number. Since the antiderivative is evaluated at the endpoints and subtracted, any constant of integration would cancel out in a definite integral.

If you can accept that there is a constant of integration in the case of an indefinite integral, we may proceed to try to understand this constant. We sometimes refer to it as an 'arbitrary' constant of integration. When we just look at the indefinite integral without any context, the constant does seem arbitrary. How can you know if it's zero, positive seven, negative three, one hundred, or anything else? You can't. But with context you can.

We'll take a tour of a variety of real-world examples of integrals in the following chapter. When a real physical quantity is represented by an integral, it can't be arbitrary. In the natural universe, if you measure a quantity, you get a definite[95] value that can be calculated using a formula. There is no guessing about a constant to add to it. To help see how a real physical quantity relates to a constant of integration, consider the following example.

Suppose that a car travels with a velocity given by the function $v_x(t) = t^2/12$. (The symbol v_x, where x is a subscript, represents the velocity of the car. We discussed this notation in Footnote 94.) Suppose that we also know that the car's position is $x = 8$ meters when $t = 0$. We can use calculus to find out where the car is when $t = 6$ seconds.

Velocity is a derivative of position with respect to time: $v_x = dx/dt$. This means that position[96] is the integral of velocity with respect to time. To find the antiderivative velocity, compare $t^2/12$ with the general form[97] at^b to identify[98] $a = 1/12$ and $b = 2$.

Now use the antiderivative formula for polynomial terms to get $\dfrac{at^{b+1}}{b+1} + c = \dfrac{\frac{1}{12}t^{2+1}}{2+1} + c =$

---

[95] Well, some physical quantities, like rolling a six-sided die or like the wave function in quantum mechanics, depend on probabilities. But while a single event may be subject to chance, when the event is repeated a large number of times, then the outcome becomes fairly well determined. Roll a die once and it may be 1, 5, 3, 2, 6, or 4, with no way of knowing which. But roll it 6,000,000 times and we can be fairly certain that the number of times it shows a 5 will be in the neighborhood of a million. If a photon passes through a narrow slit, you can't know where it will show on a screen, but if you pass a billion photons through a slit, you can predict with incredible accuracy what the interference pattern will look like.

[96] It's more precise to say that net displacement is the integral of velocity with respect to time. Net displacement is the change in position, which is a straight line from the initial position to the final position. But saying that position is the integral of velocity with respect to time will lead to the same result via the constant of integration.

[97] Here, the independent variable is time, while x is the dependent variable. If we graph x as a function of t, we would put t on the horizontal axis and x on the vertical axis. This should seem strange to you the first time you see it. But we discussed this very issue in Chapter 9. After seeing it enough times, it should start to seem less strange. If it still seems strange to you, it might be worth reviewing Chapter 9. Since $v_x = dx/dt$ is a derivative with respect to t, we need to integrate v_x with respect to time in order to find x. In this problem, $v_x = t^2/12$ is a function of time and we're integrating with respect to t (not x), which is why we wrote at^b (instead of ax^b).

[98] Note that $t^2/12$ could alternatively be expressed as $(1/12)t^2$. Whether you divide by 12 or multiply by 1/12, you get the same result. For example, note that $36/12 = 3$ is the same as $(1/12)36 = 3$. Either way you write it, one-twelfth of thirty-six equals three. If you're struggling to identify a and b with $t^2/12$, then it might help to rewrite this as $(1/12)t^2$.

$(1/12)t^3/3 + c = t^3/36 + c$. (When you divide by 12 and also divide by 3, this equates to dividing by 36. We're multiplying the fractions 1/12 and 1/3 to make 1/36.) Here, we included the important constant of integration, c, with our answer. The position of the car at time t is given by the expression $x(t) = t^3/36 + c$.

$$x = \int v_x \, dt = \int \frac{t^2}{12} \, dt = \frac{t^3}{36} + c$$

Oh, no! There is a "+ c" in our answer. Does that mean that the car's position is arbitrary? Of course not. In reality, cars don't wind up in arbitrary positions. The car's position is determined by its velocity (which is determined by its acceleration). There is a definite answer for where the car is at any given time. Now we'll see how to figure this out for this example.

We just need a little information to figure out what the value of c is, and if you reread the complete problem, you may discover that there is some information that we haven't yet utilized. We were told that the car's position was $x = 8$ meters at t = 0. We were given that $x(0) = 8$; that is, if we plug t = 0 into $x(t)$, we get $x = 8$. So, we'll take our equation from the indefinite integral, $x(t) = t^3/36 + c$, and replace t with 0 and replace x with 8. This gives us $8 = 0^3/36 + c$, which simplifies to $8 = 0 + c$ or $8 = c$. Using the information that $x = 8$ at t = 0, we were able to determine that the constant of integration was c = 8. In this example, c has a very natural physical interpretation. It is the initial position of the car (that is, where the car is when t = 0).

Are you wondering if it was reasonable to know this additional information? This information really isn't extraneous. If you see a car moving and would like to predict where the car will be 8 seconds from now, obviously to do that you need to know where the car is at some time. If you know where the car is now (for example) and if you also know the velocity function $v_x(t)$, you can predict where the car is later. If you don't have any idea where the car is now and also have no idea where the car is at any time ever, how can you expect to predict where it will be later? You need to know something about the car's motion (where it is now, or where it is when t = 3 seconds, for example) in order to be able to predict where it will be at some other time. This is the role of the constant of integration. It allows us to use relevant information (like where an object is in the beginning of the problem) to make a prediction (like where the object will be at some later time).

The example asked us to find where the car is at t = 6 seconds. The equation for the position of the car is $x(t) = t^3/36 + 8$ (because we found that c = 8). Plug in t = 6 to find the position of the car at this time: $x(6) = 6^3/36 + 8 = 216/36 + 8 = 6 + 8 = 14$

meters. (Obviously, this car wasn't driving very fast during this period; it only advanced from $x = 8$ to $x = 14$ meters in 6 seconds. Maybe it was driving in a parking lot during this time interval.)

Let's look at another example. Suppose that a rocket has a total mass of 3000 kg at t = 0, which includes a payload of 500 kg (which is the mass of the rocket without any fuel) plus 2500 kg of fuel initially. Suppose also that the rocket burns fuel at a constant rate of 40 kg/s (kilograms per second), which is referred to as the burn rate. The formula for the burn rate is $R_b = -dm/dt$, where R_b is the burn rate and m is the total mass of the rocket at time t. Since the mass m of the rocket is decreasing (since it is burning fuel to eject gases), dm/dt is negative; the reason for the minus sign in R_b = –dm/dt is to cancel the minus sign associated with dm/dt (to make R_b positive). We'd like to know the ratio of the total mass of the rocket at t = 50 seconds to the total mass of the rocket initially. (Are you wondering how you're supposed to solve this problem when you don't know anything about rockets? The truth is that you don't need to know anything about rockets. If you'd never heard the terms payload or burn rate, it doesn't really matter, since they were defined in the paragraph. And the only equation you need to know was stated in the problem. There is actually enough information here to reason out the answer without even doing any calculus.[99] But we still wish to do the calculus in order to show another example of a constant of integration.)

We begin with the equation $R_b = -dm/dt$. The burn rate equals the negative of the derivative of the mass with respect to time. If we knew an equation for mass as a function of time, we would take a derivative. But we don't. What we do know is the burn rate, R_b; it's a constant equal to 40 kg/s. How do we find the mass of the rocket at time t using the burn rate? It's the opposite of a derivative. It's the antiderivative. We will integrate $-R_b$ with respect to time (preserving the minus sign from the derivative that were we given in the problem).

$$m = -\int R_b \, dt = -\int 40 \, dt$$

This integral is easy because the burn rate is constant: $R_b = -40$. What's the antiderivative of a constant? The antiderivative of -40 is equal to $-40t + c$, where c is

---

[99] Burning fuel at a constant rate of 40 kg/s for 50 seconds, the amount of fuel burned will be (40)(50) = 2000 kg. Since the rocket started with 2500 kg of fuel, at t = 50 seconds, there will be 2500 – 2000 = 500 kg of fuel left. Add that to the 500 kg payload to find that the total final mass of the rocket is 500 + 500 = 1000 kg. Divide 1000 kg by the total initial mass of the rocket, 3000 kg, to find that the specified ratio is 1/3.

a constant of integration.[100] (It's easy to check. If you take a derivative of $-40t + c$ with respect to t, you get $-40 + 0 = -40$.) Thus, our formula for the mass of the rocket at time t is $m(t) = -40t + c$.

How do we find the constant of integration? Again, we look at information given in the problem that we haven't already used. The rocket has a total mass of 3000 kg at $t = 0$. Replace m with 3000 and t with 0 in the equation for mass. This gives $3000 = -40(0) + c$, which simplifies to $3000 = -0 + c$, which means that $3000 = c$. In this example, the constant of integration is the initial total mass of the rocket. With $c = 3000$, our equation for the mass becomes $m(t) = -40t + 3000$.

The problem wants to know about the mass at $t = 50$ seconds, so plug this value for t into the equation for mass: $m(50) = -40(50) + 3000 = -2000 + 3000 = 1000$ kg. (This already includes the 500 kg payload.) Divide the final total mass of 1000 kg by the total initial mass of 3000 kg to get the ratio that the problem asked for: $1000/3000 = 1/3$.

What did we learn in this chapter? In a real problem, we can use available information (like the initial value of a quantity) to solve for the constant of integration. It's common for the constant of integration to have a natural, physical interpretation. In these two examples, one constant of integration represented the initial position of the car and the other represented the total initial mass of the rocket.

Quick Check (Ch. 15)

Do you understand the main idea from Chapter 15?

1. Why does the answer to an indefinite integral include a constant of integration?

---

[100] Note that $-40 = -40t^0$ since $t^0 = 1$ (recall Footnote 32). If you're trying to use the formula for the antiderivative of polynomial terms, $a = -40$ and $b = 0$, which gives you $\frac{at^{b+1}}{b+1} + c = -\frac{40t^{0+1}}{0+1} + c = -40t^1/1 + c = -40t + c$.

16 Which physical quantities involve integrals?

Integrals are very common in many applications of calculus, including science, engineering, and even other subjects like economics. In this chapter, we'll explore a variety of examples of physical quantities that involve integrals.

Recall from Chapter 9 that velocity is a derivative of position with respect to time. Since an indefinite integral is basically an antiderivative, this means that the integral of velocity with respect to time tells us about the object's position. More precisely, the definite integral of velocity with respect to time equals the **net displacement**, which is a straight line from the initial position to the final position. The net displacement tells you how the position of the object has changed.

It's interesting to note that integrating speed may result in a much different answer than integrating velocity. Do you remember what the distinction is between speed and velocity? (We mentioned this in a previous chapter.) Velocity is a combination of speed and direction. The statement "the car is traveling 30 m/s" gives the speed of the car, but not its velocity, since it tells how fast the car is moving with no mention of which way the car is headed. In contrast, the statement "the car is traveling 30 m/s northeast" gives the velocity of the car since it tells both how fast the car is moving and which way it is moving. The only difference between speed and velocity is that velocity includes direction. Yet what may seem like a subtle distinction can make a big difference. Whereas the definite integral of velocity with respect to time is the net displacement of the object, the definite integral of speed with respect to time is the **total distance traveled**.

We can illustrate this distinction without even doing the calculus. Suppose that a jogger runs exactly halfway around a circular track. If you were given the jogger's velocity and speed, each as a function of time, you could carry out the integral to find the net displacement and total distance traveled. We won't do these integrals, but we can use the meaning of net displacement and total distance traveled to understand what the answers to those integrals would be. If you integrate the velocity with respect to time, you get the net displacement. Do you remember how we defined net displacement? It's a straight line from the jogger's initial position to the jogger's final position. Draw a circle, mark one point on the circle as the initial position, and mark the final position. What do you call the line that connects the initial and final positions in this example? A diameter. If the jogger runs exactly halfway around a circular track, the final position and initial position will be separated by one diameter. In contrast, what is the total distance traveled? The total distance traveled will be one-half of the circumference of the circle. (Recall that circumference is the total distance around a circle.) In this example, the total distance traveled is a bigger distance than the magnitude of the net displacement.

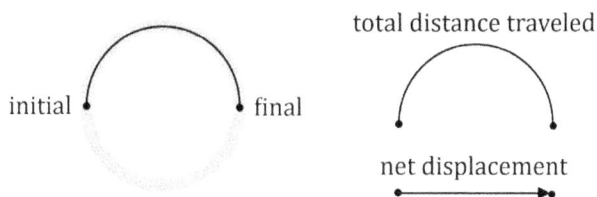

Calculus students call the total distance traveled by another name: **arc length**. It isn't necessary to know anything about an object's speed to find the arc length. You just need to know the curve that the object travels along (and the points where it starts and finishes). The arc length integral challenges students in two ways. First, it looks a bit strange and intimidating the first time that you see it. Second, it involves a square root that makes the technique of performing the integral more difficult.

We'll set up the arc length integral to show you how it looks so that you can appreciate the first point, but there is no need to worry, since we won't actually do the integration. Before we write down the arc length integral, we'll begin with a few trivial cases. First, suppose that an object travels in a straight line in the positive x-direction, starting at x_1 and finishing at x_2. In this simple case, the differential arc length is simply the differential dx and the arc length is simply the integral of dx. The integrand in this case is one, which couldn't be simpler. Since the antiderivative of one is x (because a derivative of x with respect to x equals one), the definite integral gives $x_2 - x_1$, which is just the change in the x-coordinates. Similarly, if an object travels in a straight line along positive y, we integrate dy to find that the arc length is $y_2 - y_1$. (In the simple case of traveling in a straight line, the arc length and the net displacement are the same distance.)

In the more general case of (curved) motion within the xy plane, the arc length is the square root of $dx^2 + dy^2$. This comes from using the Pythagorean theorem at the infinitesimal level. (The Pythagorean theorem doesn't apply at the finite level if the path is curved because the Pythagorean theorem applies to triangles, not curves. But for an infinitesimal displacement, it works.) To find the arc length, we integrate the square root of $dx^2 + dy^2$. How do you do that? Good question. The trick is to factor out dx^2 to write $dx^2 + dy^2 = [1 + (dy/dx)^2]dx^2$. Since the square root of dx^2 is simply dx, the arc length is the integral of the square root of $[1 + (dy/dx)^2]\,dx$, as shown below. In a typical problem, the student would be given the equation of the curve. For example, if an object travels along the parabola $y = x^2$, a student could use this function to determine the derivative $dy/dx = 2x$ in this case, then replace dy/dx with $2x$ in the arc length integral. This gives the square root of $[1 + (2x)^2]$, which simplifies to the square root of $[1 + 4x^2]$, and then the student needs to be fluent in techniques of integration to finally find the arc length.

$$\text{arc length} = \int_i^f \sqrt{dx^2 + dy^2} = \int_i^f \sqrt{1 + \left(\frac{dy}{dx}\right)^2} \, dx$$

In contrast, finding net displacement is a trivial matter. If you know the coordinates (x_1, y_1) and (x_2, y_2) where the object begins and ends, you can simply use the distance formula[101] from geometry to find the magnitude[102] of the net displacement. We don't even need calculus for this. Since net displacement is a straight line from (x_1, y_1) to (x_2, y_2), this problem is much simpler than arc length (when an object travels along a curved path).

Recall from Chapter 9 that acceleration is a derivative of velocity with respect to time. Since an indefinite integral is basically an antiderivative, this means that the integral of acceleration with respect to time tells us about the object's velocity. More precisely, the definite integral of acceleration with respect to time equals the **change in velocity** (which is the final velocity minus[103] the initial velocity).

If an object is displaced along the $+x$-axis, the **work** done by a particular force is the definite integral of the x-component of the force with respect to x. Work integrals are important because work is related to energy. In particular, according to the work-energy theorem, the net work (which is the work done by the net force, meaning the vector sum of the forces acting on an object) equals the change in the kinetic energy of the object. (Kinetic energy is defined as work that can be done by changing an object's speed; it is the energy of motion.)

Another area of physics and engineering where integrals are very common is electromagnetism. The **electric field** (which is force per unit charge) due to a conti-nuous distribution of electric charge (such as a uniformly charged rod or sphere) is found by an integral.[104] The **magnetic field** due to a stream of moving charges (also

---

[101] Find $(x_2 - x_1)^2 + (y_2 - y_1)^2$ and then take the square root. The distance formula is basically just the Pythagorean theorem. Given (x_1, y_1) to (x_2, y_2), you can form a right triangle where the base is $x_2 - x_1$, the height is $y_2 - y_1$, and the hypotenuse is the net displacement.

[102] The word 'magnitude' just means 'how much.' Net displacement is a vector; it has a magnitude and a direction. The distance formula gives you the magnitude. If you also want the direction, you need to apply trigonometry. The direction is an angle; it's the inverse tangent of $(y_2 - y_1)/(x_2 - x_1)$.

[103] Since velocity is a vector, when the object takes a curved path, this is actually a vector subtraction problem. Trigonometry students use their trig skills to add or subtract vectors.

[104] Physics students learn two different integrals for finding electric fields. One is a direct integral involving dq/R^2, where q represents electric charge and R is the distance from dq to

known as current) is similarly found by an integral.[105] Maxwell's equation (Chapter 21) are often expressed in integral form as four different integrals (involving vector calculus).

In quantum mechanics, the probability of finding a particle in a particular region of space can be found by an integral involving the **wave function**. One first solves Schrödinger's equation (Chapter 21) to determine the wave function, and then one uses the wave function to calculate probabilities or expectation values. Specifically, probability is the definite integral of psi-star psi, where psi is the Greek letter ψ representing the wave function and psi-star is its complex conjugate (where the word 'complex' refers to the subject of mathematics involving complex numbers).

Integrals are common in probability (for systems with continuous rather than discrete variables) and statistics. For example, the normal distribution taught in statistics courses involves the integral of the Gaussian function, which is an exponential function of the form $\exp[-a(x-b)^2]$, where exp represents the exponential function (recall Chapter 5[106]). The normal distribution is important because it applies to systems that are subject to random chance. Measurements have inherent errors because you can't measure physical quantities exactly (except in the special case that you're counting a small number of objects, like the number of tires on a semi-truck). When measurements have random errors (meaning that the measurement is just as likely to be a little too high as it is to be a little too small), the normal distribution applies. (In contrast, systematic errors cause the measurements to be skewed one way or another, like the way that air resistance causes the speed of an object to be smaller than otherwise expected.)

While integrals are especially common in science, engineering, and pure mathematics, they are also encountered in other subjects. For example, in economics, a company's revenue can be found by integrating marginal revenue with respect to the amount of product manufactured.

Finally, any relationship that can be expressed as a derivative can also be expressed as an integral. So, all of the examples of physical quantities that are derivatives mentioned in Chapter 9 can also be expressed as integrals. For example, electric current is a derivative of charge with respect to time. Therefore, charge is the definite integral of current with respect to time.

---

the field point (which is the point where you're trying to find the electric field). The other integral is known as Gauss's law.

[105] Physics students know these integrals as the law of Biot-Savart and Ampère's law.

[106] Back in Chapter 5, we introduced the simple exponential function e^x, which could alternatively be expressed as $\exp(x)$.

Pop Quiz (Ch. 16)

See if you remember these points from Chapter 16.

1. What does the definite integral of velocity with respect to time give you?

2. What does the definite integral of acceleration with respect to time give you?

17 What are Rolle's theorem and the mean value theorem?

In Chapter 14, we learned about the first and second fundamental theorems of calculus. In this chapter, we will explore two more significant theorems that pertain to calculus: Rolle's theorem and the mean value theorem.

First, we will consider Rolle's theorem. Suppose that a function $f(x)$ equals zero when x = A and also equals zero when x = B, meaning that $f(A) = f(B) = 0$. This is easy to visualize if we graph $f(x)$. For such a function, the curve crosses the x-axis at the points x = A and x = B, like the example shown below. According to **Rolle's theorem**, for such a function, if $f(x)$ is both continuous and differentiable over the interval (from A to B), then there must be at least one point between A and B for which the derivative $\frac{df}{dx}$ is equal to zero.

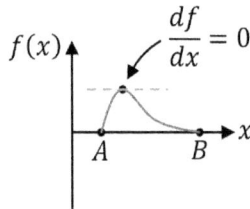

To help understand Rolle's theorem, imagine a bug crawling along $f(x)$ from x = A to x = B. Since $f(A) = 0$ and $f(B) = 0$, the bug's journey begins and ends on the x-axis. The bug begins on the x-axis at point A and will return to the x-axis at point B. Consider the cases below.

• If $\frac{df}{dx}$ is positive at x = A, then the slope is positive at x = A. In this case, the bug begins its journey by traveling above the x-axis, like the graph below on the left. In order for the bug to return to the x-axis at x = B, it must travel downward with negative slope for some region between A and B. If the slope is positive to begin with and negative at some time before reaching B, then the slope will be horizontal (at least momentarily) when it changes from positive to negative. (We know this because the function was said to be both continuous and differentiable, meaning that the curve is smooth without any gaps.) When the slope is horizontal, the value of the slope is zero, and $\frac{df}{dx}$ is zero at that point.

• If $\frac{df}{dx}$ is negative at x = A, then the slope is negative at x = A. In this case, the bug begins its journey by traveling below the x-axis, like the graph below on the right. In order for the bug to return to the x-axis at x = B, it must travel upward with positive slope for some region between A and B. If the slope is negative to begin with and positive at some time before reaching B, then the slope will be horizontal (at least momentarily) when it changes from negative to positive. (As in the previous bullet point, we can draw this conclusion since the function was said to be continuous and smooth.) When the slope is horizontal, the value of the slope is zero, and $\frac{df}{dx}$ is zero at that point.

• If $\frac{df}{dx}$ is zero at x = A, then the slope is horizontal at x = A. If the slope remains horizontal all the way to x = B, then $\frac{df}{dx}$ is zero for every point from A to B, which satisfies Rolle's theorem. Otherwise, the slope will turn positive or negative before reaching B, and then one of the two previous bullet points demonstrates that $\frac{df}{dx}$ is zero somewhere between A and B.

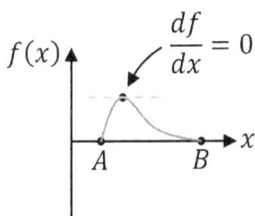

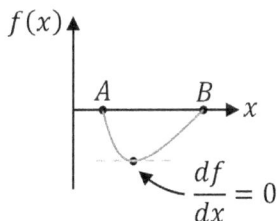

Note that in Rolle's theorem f(x) is both continuous and differentiable from A to B. Because it is differentiable, the curve is smooth. These conditions are important for the theorem. If f(x) didn't have to be continuous, it would be possible to reach point B without having a horizontal slope, like the example below on the left. If f(x) didn't have to be differentiable, it would again be possible to reach point B without having a horizontal slope, like the example below on the right. (At the sharp peak below on the right, the tangent and thus the derivative are not defined. This curve is not smooth and therefore is not differentiable.) If f(x) is both continuous and differentiable (and therefore smooth), then there will be at least one point with a horizontal tangent (where $\frac{df}{dx} = 0$) between A and B.

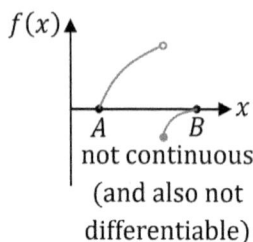

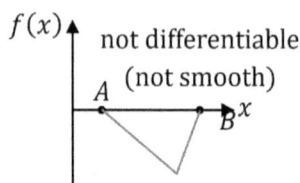

What if the function isn't zero at A, B, or both points? In that case, we get a more general theorem called the mean value theorem. According to the **mean value theorem**, if f(x) is both continuous and differentiable over the interval from A to B, then there must be at least one point between A and B for which the derivative $\frac{df}{dx}$ is equal to the slope of the line that joins points A and B. For example, consider the function f(x) illustrated below on the left. In this example, neither f(A) nor f(B) is zero. Furthermore, f(A) isn't even equal to f(B); in this case, f(A) is smaller than f(B). Compare this with the diagram below on the right, which is a similar graph corresponding to Rolle's theorem. We drew the graph below on the right by rotating (and shifting) the graph on the left. Both graphs below illustrate the same fundamental principle; if f(x) is continuous and differentiable from A to B, then at some point between A and B the derivative $\frac{df}{dx}$ will have the same slope as the line joining A and B. What's special for Rolle's theorem is that the line joining A and B is horizontal, which has zero slope; thus $\frac{df}{dx}$ equals zero for some point between A and B. The mean value theorem states the same idea more generally, that some point between A and B will have a tangent line with the same slope as the line that joins A and B. (By the way, the line that joins A and B is called a secant.[107])

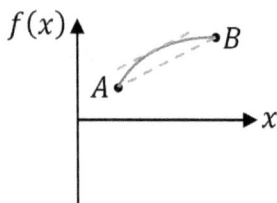

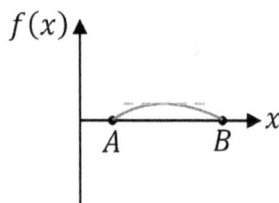

---

[107] To clarify, the infinite line that passes through both points A and B is called a secant, whereas the finite line segment beginning at A and ending at B is called a chord. Students first learn the terms 'chord' and 'secant' in geometry courses when they study theorems concerning circles.

Recall from Chapter 8 that the slope of a line can be calculated by dividing the difference in y-values by the corresponding difference in x-values. Since we plotted f on the vertical axis, in this context we'll divide a difference in f-values by the corresponding difference in x-values. To find the slope of the line that joins points A and B, the f-values are f(B) and f(A). The slope of the line that joins A to B is equal to $\frac{f(B)-f(A)}{B-A}$. Thus, the mean value theorem states that if f(x) is continuous and differentiable from A to B, then at some point between A and B the derivative $\frac{df}{dx}$ is equal to $\frac{f(B)-f(A)}{B-A}$. This formula equates to saying that there exists a point between A and B where the tangent line has the same slope as the line that joins A and B.

The mean value theorem has a few important corollaries, which are mentioned below. The first corollary below may seem trivial, but it is used to obtain the corollaries below it.

- In the simple case that $\frac{df}{dx}$ equals zero for every point from A to B, then f(x) equals a constant. A graph of f(x) for this case is a horizontal line.
- If g(x) and h(x) are two different functions and their derivatives $\frac{dg}{dx}$ and $\frac{dh}{dx}$ are equal over the interval from A to B, then g(x) and h(x) differ only by a constant. This point follows from the previous point. To see this, let f(x) = g(x) – h(x). If we take a derivative of both sides, we get $\frac{df}{dx} = \frac{dg}{dx} - \frac{dh}{dx}$. Since $\frac{dg}{dx} = \frac{dh}{dx}$ in this case, this shows that $\frac{df}{dx}$ = 0 for every value of x from A to B. According to the previous bullet point, f(x) equals a constant, which means that g(x) – h(x) equals a constant.
- The previous bullet point explains why the most general form of an indefinite integral is the antiderivative of the integrand plus a constant. In the previous bullet point, g(x) and h(x) are two different antiderivatives of $\frac{dg}{dx}$ (since $\frac{dg}{dx}$ and $\frac{dh}{dx}$ are equal), and g(x) and h(x) can differ by no more than a constant.

Note that there is another mean value theorem which pertains to definite integrals. According to the **mean value theorem of definite integrals**, if f(x) is continuous over the interval from A to B, then there exists a point C between A and B such that the definite integral of f(x) dx from x = A to B is equal to f(C) times (B – A). The value of f(C) is interpreted as the **average value of the function over this interval**.

Challenge Question (Ch. 17)

1. In which way is Rolle's theorem a special case of the mean value theorem?

18 How does calculus help to find volume?

We've already seen some examples of how calculus is helpful in geometry. For example, we've seen that derivatives tell us about the slopes of tangent lines (Chapter 8), second derivatives tell us about the concavity of a curve (Chapter 11), definite integrals tell us about the area between a curve and the horizontal axis (Chapter 14), and the arc length integral tells us the length along a curve (Chapter 16).

One area of geometry where calculus is particularly helpful is with three-dimensional objects that have curved surfaces like an ellipsoid[108] or a donut.[109] The formulas for the volume or surface area of such objects are found using integrals. Calculus teaches a few different methods for calculating the volume or surface area of an object with a curved surface, including the method of slicing, the method of revolution, and performing double or triple integrals.

For the **method of slicing**, we imagine slicing a three-dimensional solid much like a chef would slice a fruit or vegetable. To use the method of slicing, set up a coordinate system where the x-axis is perpendicular to the slices, like the example below. Next, express the area of each slice as a function of x. Finally, integrate the area $A(x) \, dx$. (Why? The integral with respect to x effectively adds up the areas of all the slices, like the Riemann sum that we discussed in Chapter 14, to make the volume.)

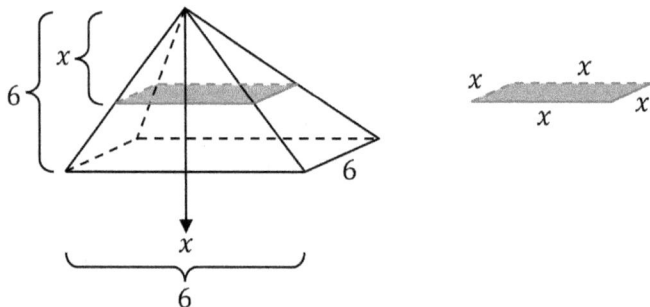

As an example, we will use this method to find the volume of the square pyramid illustrated above. This particular pyramid is special; it has a square base with an edge

---

[108] An ellipsoid is similar to a sphere, except that it is either stretched or compressed in one direction, much like the distinction between an ellipse and a circle.

[109] A single-holed ring torus has the shape of a typical donut. The next time you buy donuts, try asking for a dozen single-holed ring tori with sprinkles. If someone spends their days baking or selling donuts, that person ought to know the proper term for the shape, right? Or maybe not...

length of 6 and it also has a height of 6. The slices in this example are squares. If the distance of the slice from the origin is x, the edge length of the square slice is also x.[110] The area of the slice is $A(x) = x^2$ (since the area of a square is found by squaring the edge length). The volume of the pyramid is the definite integral of $x^2\, dx$ from $x = 0$ to $x = 6$. Do you remember how to find the antiderivative of a polynomial expression? We learned this in Chapter 14. Compare ax^b with x^2 to see that a $= 1$ and b $= 2$. Use the formula $\dfrac{ax^{b+1}}{b+1}$ to get $\dfrac{1x^{2+!}}{2+1} = \dfrac{x^3}{3}$. (It's easy to check. Take a derivative of $x^3/3$ with respect to x. Since this derivative equals the integrand, x^2, our answer checks out.) To find the definite integral, evaluate $x^3/3$ at $x = 6$, evaluate $x^3/3$ at $x = 0$, and subtract. When $x = 6$, we get $6^3/3 = 216/3 = 72$. When $x = 0$, we get $0^3/3 = 0$. Subtraction gives $72 - 0 = 72$. This agrees with the formula for the volume of a pyramid with a square base: V = base area times height divided by three. The base area is $6^2 = 36$ and the height is 6, such that V $= 36(6)/3 = 216/3 = 72$.

The **method of revolution** can be used when the shape can be formed by drawing a region in a plane and then rotating the region. The example below illustrates how a triangle can be rotated in a circle so as to form a cone. The axis of rotation is the height of the triangle, which is also the height of the cone. To use the method of revolution, set up a coordinate system with the x-axis as the axis of rotation, like the example below. Next, consider a value of x that is representative of how the solid is formed by revolution. Look at the plane region that is rotated to form the solid. For that value of x, draw a line segment that is perpendicular to the x-axis. This line segment is the radius of a circle that forms during the rotation, as indicated below. Express the radius R as a function of x. Finally, integrate $\pi[R(x)]^2\, dx$, where π is the lowercase Greek letter pi, representing the number that begins 3.14159265... (Why? Because π times radius-squared is the area of each circle. The integral with respect to x effectively adds up, like the Riemann sum that we discussed in Chapter 14, all of the areas for each value of x.) Calculus students can use a similar method to find surface area. For surface area, the integrand is instead $2\pi\, R(x)$, where 2π times radius is the circumference of each circle. The circumference of the circle is the part of the circle lying on the surface of a solid of revolution.

---

[110] This worked out nicely because the height and edge length both equal 6. If the height were different, the problem would be more complicated; we'd need to use the equation of a straight line to relate the base to the height in order to express the edge length of the slice in terms of x. That's one of the kinds of challenges that calculus students face. Our example is simple so that you can focus on the main idea.

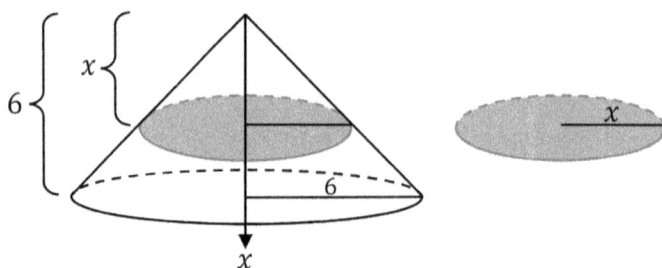

As an example, we will use this method to find the volume of the right-circular cone illustrated above. This particular cone is special; it has a circular base with a radius of 6 and it also has a height of 6. This example is easy since it turns out that $R(x) = x$; that is, the radius of each circle is equal to x.[111] Since $R(x) = x$, we need to integrate πx^2 dx. This is just like our previous example, except for the constant π. Recall from the previous example that the antiderivative of x^2 is $x^3/3$. The indefinite integral here is thus $\pi x^3/3$. For the definite integral, evaluate this at $x = 6$ and $x = 0$, and subtract. When $x = 6$, we get $\pi 6^3/3 = 216\pi/3 = 72$ pi. When $x = 0$, we get $\pi 0^3/3 = 0$. Subtraction gives $72\pi - 0 = 72\pi$. This agrees with the formula for the volume of a right-circular cone: V = base area times height divided by three. The base area is $\pi 6^2$ $= 36\pi$ and the height is 6, such that V = 36π times 6 divided by 3 = 36π times 2 = 72π, which is approximately equal to 226.195. (This particular example could have been solved by using the method of slicing or the method of revolution. Sometimes, one choice seems better than the other; other times, neither method is suitable.) Our answer agrees with the formula V = $\pi R^2 H/3$.

A more general method for finding volume is to perform a triple integral, and a more general method for finding surface area is to perform a double integral. These methods are taught in multivariable calculus. We'll briefly describe what multivariable calculus is in Chapter 20.

Ready for a difficult volume integral? (Ch. 18)

If you finished reading Chapter 18, you get an A. (Performing a volume integral is way beyond the scope of this book. It's enough just to read about it.)

---

[111] This worked out nicely because the height and radius both equal 6. If the height were different, the problem would be more complicated; we'd need to use the equation of a straight line to relate the radius to the height in order to express the radius of the circle in terms of x. As with the previous example, we spared you this complication in order to help illustrate the main idea.

19 What are sequences and series?

Up until now, we've discussed limits, derivatives, and integrals. Many calculus students would tell you that these are the three main topics that they learn in a calculus course. Infinite sequences and series are another important topic that calculus students learn about.

First, it's important to distinguish between the terms 'sequence' and 'series.' A **sequence** is a list of elements (which are usually numbers) in a particular order. For example, $\frac{1}{2}, \frac{1}{3}, \frac{1}{5}, \frac{1}{7}, \frac{1}{11}, \frac{1}{13}, \frac{1}{17}, \frac{1}{19}, \frac{1}{23}, \frac{1}{29}, \ldots$ is a sequence of fractions where the numerator is always one and the denominator is the next prime number. The first 10 elements of this sequence were given, and the three dots (…), which are called an **ellipsis**, indicate that the sequence continues forever. In a **series**, the elements of the sequence are added together. For example, $\frac{1}{2} + \frac{1}{3} + \frac{1}{5} + \frac{1}{7} + \frac{1}{11} + \frac{1}{13} + \frac{1}{17} + \frac{1}{19} + \frac{1}{23} + \frac{1}{29} + \cdots$ is the series corresponding to the sequence from the previous example.

It's really important to pay attention to whether we are talking about a sequence or a series. For example, we will briefly explore the terms convergence and divergence in this chapter. The rules for whether a sequence converges are different from the rules for whether a series converges. So, if one person is talking about the convergence of a series, but someone else is thinking about the convergence of the corresponding sequence, such a communication problem could result in an unnecessary dispute.

If you see numbers separated by commas, as in 16, 8, 4, 2, 1, 1/2, 1/4, 1/8, …, you know that it's a sequence, and if you see numbers separated by plus signs, as in $16 + 8 + 4 + 2 + 1 + 1/2 + 1/4 + 1/8 + \cdots$, you know that it's a series. But since we sometimes use other kinds of notation for sequences and series, you'll want to learn a little basic notation so that you can tell what's a sequence and what's a series when we don't write out a list of numbers.

Following are common ways of denoting a **sequence**. Recall that a sequence is a list of numbers (which aren't being added together).

• The most obvious method is to list numbers separated by commas, like 10, 16, 22, 28, 34, 40, 46, 52, 58, 64, … When there is an ellipsis (…) at the end, the sequence is infinite.

• Another method is to enclose an expression in braces, like $\{6n + 4\}$. With this method, the variable is called an **index**. In the example $\{6n + 4\}$, the index is n (but just as in algebra, we can use other letters like k or i). The index represents an integer. It's understood that the sequence begins with $n = 1$. For the next term (each number in the sequence is called a **term**), $n = 2$, then $n = 3$, and so on. For this example, the first number in the sequence is $6(1) + 4 = 6 + 4 = 10$, the second number is $6(2) + 4 = 12 + 4 = 16$, the third number is $6(3) + 4 = 18 + 4 = 22$, etc. The sequence denoted by $\{6n + 4\}$ is equivalent to the sequence in the previous bullet point, since it generates the same list of numbers in the same order. If you see the expression $6n + 4$ without the braces, this only refers to the term for a specific value of n, whereas the expression $\{6n + 4\}$ with the braces refers to the entire sequence. With this notation, a sequence can alternatively be defined as a function (recall Chapter 5) where the argument is a positive integer.

• A sequence can be referred to using a letter and a subscript, as in $\{a_n\}$. Once again, the braces indicate that the terms a_n form a sequence. If we write a_n without braces, this means just the one term for a specific value of n, whereas $\{a_n\}$ with braces, this refers to the entire sequence. When using this notation, one must also provide a formula for a_n. If we let $a_n = 6n + 4$, then the sequence $\{a_n\}$ is identical to the sequence in the previous bullet points. One reason for using this notation is that it allows us to refer to two different sequences by name. For example, if $b_n = \dfrac{1}{n}$ and $c_n = n^2$, then we can make statements like $\{b_n\}$ converges to zero whereas $\{b_n\}$ diverges. (This statement will make more sense once we discuss what it means for a sequence to converge or diverge.) The point of this notation is that the a's, b's, and c's help us refer to different sequences.

• A recursion relation is yet another common method. A **recursion relation** tells you how two (or more) consecutive terms are related; it also specifies an initial value. An example of a recursion relation is $a_{n+1} = a_n + 6$ with $a_1 = 10$. The initial term is $a_1 = 10$. The second term is $a_2 = a_1 + 6 = 10 + 6 = 16$. The third term is $a_3 = a_2 + 6 = 16 + 6 = 22$. The next term is $a_4 = a_3 + 6 = 22 + 6 = 28$. Observe that this recursion relation produces the same sequence as the previous three bullet points.

Following are common ways of denoting a **series**. Recall that in a series the numbers are added together.

• The most obvious method is to separate numbers with plus signs like $9 + 3 + 1 + 1/3 + 1/9 + 1/27 + 1/81 + 1/243 + \cdots$ Again, the ellipsis ($\cdots$) indicates that the series is infinite, meaning that the series continues indefinitely; there are an infinite number of terms.

• Sometimes a formula with an index follows a sum of numbers, like $9 + 3 + 1 + 1/3 + 1/9 + 1/27 + 1/81 + 1/243 + \cdots + 27/3^n + \cdots$ Here, $27/3^n$ indicates that each term of the series can be calculated using the formula $27/3^n$. When $n = 1$ we get $27/3^1 = 27/3 = 9$, when $n = 2$ we get $27/3^2 = 27/9 = 3$, when $n = 3$ we get $27/3^3 = 27/27 = 1$, when $n = 4$ we get $27/3^4 = 27/81$ which reduces[112] to 1/3, etc.

• More commonly, when the formula for a series is given, it is done using **summation notation** like the example below. The large symbol at the left of the example below is the uppercase Greek letter sigma (so if it appears Greek to you, that's partly because Σ actually is Greek); it is called the **summation symbol**. This sigma symbol is used in calculus to indicate a series. The variable below it is referred to as the **index**; in the example below, the index is n. The starting value follows the equal sign; in the example below, n begins at 1. The upper limit appears above the sigma; in the example below, the upper 'limit' is infinity, meaning that the series continues forever (yet the sum of this infinite series is finite, but we'll discuss that concept later in this chapter). Finally, the formula for the nth term of the series appears to the right of the summation symbol; in the example below, the formula is $27/3^n$. The formula works the same way as it does for a sequence, but when the formula appears to the right of a summation symbol, it represents a series with the terms added together. The notation below refers to the same series described in the two previous bullet points.

$$\sum_{n=1}^{\infty} \frac{27}{3^n} = \frac{27}{3^1} + \frac{27}{3^2} + \frac{27}{3^3} + \frac{27}{3^4} + \cdots$$
$$= \frac{27}{3} + \frac{27}{9} + \frac{27}{27} + \frac{27}{81} + \cdots = 9 + 3 + 1 + \frac{1}{3} + \cdots$$

Following are some examples of different kinds of sequences. (For each sequence below, there exists a corresponding series where the terms are added together.)

• The **Fibonacci** sequence is 1, 1, 2, 3, 5, 8, 13, 21, 34, 55, … The recursion relation for this sequence is $a_{n+2} = a_n + a_{n+1}$ where $a_1 = 1$ and $a_2 = 1$. (This particular recursion relation needs not one, but two starting values, a_1 and a_2, because each term depends on the two previous terms, as we'll see.) For example, the third term is $a_3 = a_1 + a_2 = 1 + 1 = 2$, the fourth term is $a_4 = a_2 + a_3 = 1 + 2 = 3$, the fifth term is $a_5 = a_3 + a_4 = 2 + 3 = 5$, and the next term is $a_6 = a_4 + a_5 = 3 + 5 = 8$. The

---

[112] The greatest common factor of 27 and 81 is the number 27, since 27 is the largest integer that evenly divides into both 27 and 81. Divide 27 and 81 each by 27 to see that 27/81 reduces to 1/3. Alternatively, enter 27/81 and 1/3 each on a calculator to see that each is equivalent to the decimal 0.3333333333 with the 3's repeating indefinitely.

idea is simple; add the two previous values to find the next value. For example, after 5 and 8, the next number is 5 + 8 = 13, and after 8 and 13, the next number is 8 + 13 = 21.

• The **harmonic** sequence is $\left\{\frac{1}{n}\right\}$, which makes the sequence $1, \frac{1}{2}, \frac{1}{3}, \frac{1}{4}, \frac{1}{5}, \frac{1}{6}, \frac{1}{7}, \frac{1}{8}, \ldots$ The corresponding series where these fractions are added together is particularly important in calculus, as we'll discuss later in this chapter.

• **Triangular** numbers are formed by adding together consecutive integers beginning with one. This forms the sequence 1, 1+2, 1+2+3, 1+2+3+4, 1+2+3+4+5, 1+2+3+4+5+6, …, which simplifies to 1, 3, 6, 10, 15, 21, … These are called triangular numbers because they can be formed by stacking objects in triangles as shown below. (But maybe the diagrams below seem more like pyramids to you.) It turns out that there is a simple formula for the sum of the integers $1+2+3+4+5+6+\cdots+n$, which we'll discuss later in this chapter.

1

1 + 2 = 3

1 + 2 + 3 = 6

1 + 2 + 3 + 4 = 10

1 + 2 + 3 + 4 + 5 = 15

1 + 2 + 3 + 4 + 5 + 6 = 21

• In a **geometric** sequence, the ratio of any two consecutive terms is always the same value. For example, in the geometric sequence 2, 6, 18, 54, 162, 486, …, each number is 3 times the previous value. The third term is 3 times 6 = 18, the fourth term is 3 times 18 = 54, and the next term is 3 times 54 = 162. The recursion relation for this sequence is $a_{n+1} = 3a_n$, and the formula is $\{2 \text{ times } 3^{n-1}\}$ beginning with $n = 1$. (Note that $3^{1-1} = 3^0 = 1$, since any nonzero number raised to the power of zero equals one. See Footnote 32.) For example, when $n = 2$ we get 2 times $3^1 = (2)(3) = 6$, when $n = 3$ we get 2 times $3^2 = 2$ times $9 = 18$, and when $n = 4$ we get 2 times $3^3 = 2$ times $27 = 54$. Any geometric sequence has a formula of the form $\{a \text{ times } r^{n-1}\}$, where a and r are constants.

• In an **alternating** sequence, every other term is negative. An example of an alternating sequence is $1, -\frac{1}{3}, \frac{1}{5}, -\frac{1}{7}, \frac{1}{9}, -\frac{1}{11}, \ldots$

In series calculus, one of the main kinds of problems that students face is to determine whether a given sequence or series converges or diverges. (We will define these two terms. Soon.) The test used to determine if a sequence converges is different from the tests used to determine if a series converges. We'll discuss the convergence of sequences first and then we'll discuss the convergence of series.

If a sequence **converges**, this means that as the index grows infinite, the terms approach a finite **limiting value**. The numerical value that it approaches is called the limiting value. We'll illustrate this with a few examples. The harmonic sequence $\left\{\frac{1}{n}\right\}$ $= 1, \frac{1}{2}, \frac{1}{3}, \frac{1}{4}, \frac{1}{5}, \frac{1}{6}, \ldots$ approaches the limiting value of zero. As the index n grows to infinity, its reciprocal, $\frac{1}{n}$, becomes very small. For example, when n = one million (1,000,000), its reciprocal, $\frac{1}{1,000,000}$, is equal to 0.000001. The larger n is, the smaller $\frac{1}{n}$ is. Since the harmonic sequence approaches the finite value of zero as n goes to infinity, the harmonic sequence converges. (In contrast, the harmonic series diverges, as we'll discuss later. As mentioned previously, it's really important to pay attention to whether we say sequence or series.) As a second example, the sequence $\left\{\frac{n}{n+1}\right\}$ = $\frac{1}{2}, \frac{2}{3}, \frac{3}{4}, \frac{4}{5}, \frac{5}{6}, \frac{6}{7}, \ldots$ approaches the limiting value of one. As the index n grows to infinity, the ratio $\left\{\frac{n}{n+1}\right\}$ approaches one. For example, when n = 99, the ratio is $\frac{99}{100}$ = 0.99. The sequence $\left\{\frac{n}{n+1}\right\}$ converges to the value of one.

On the other hand, if a sequence **diverges**, this means one of two things. It could mean that as the index grows infinite, the terms grow infinite. For example, the sequence $\{n^2\}$ = 1, 4, 9, 16, 25, 36, … diverges because the terms grow larger as n grows larger. There is no limit to how large n^2 will get. Another way that a sequence can diverge is if the values remain finite, but never converge to any single value. For example, the sequence 1, 2, 1, 2, 1, 2, 1, 2, … diverges because it oscillates back and forth between 1 and 2 without ever approaching a single number.

In order to determine whether a sequence converges or diverges, calculus students apply their knowledge of **limits** (Chapters 6-7). For example, to determine whether the sequence $\left\{\frac{2n}{n+3}\right\}$ converges or diverges, calculus students examine the limit as n goes to infinity of the ratio $\frac{2n}{n+3}$. The trick here is to divide the numerator and denominator each by n to get $\frac{2}{1+\frac{3}{n}}$. As n goes to infinity, the new numerator is simply 2 and the new denominator approaches 1 (since the second term of the denominator, $\frac{3}{n}$, approaches zero as n grows infinite). The ratio $\frac{2}{1+\frac{3}{n}}$ thus approaches $\frac{2}{1}$ = 2, showing that $\left\{\frac{2n}{n+3}\right\}$

converges to 2. If you plug in $n = 100$, for example, you'll get $\frac{200}{103}$, which is approximately 1.9417. For an even larger value of n, the ratio will be even closer to 2.

Sometimes calculus students need to apply their knowledge of **derivatives** to determine whether a sequence converges or diverges. This happens when they need to apply **l'Hôpital's rule** (Chapter 13). For example, consider the sequence $\left\{\frac{\ln(n)}{n}\right\}$, which involves the natural logarithm function (Chapter 5). As n grows infinite, the function $\ln(n)$ grows to infinity and the denominator n also grows to infinity. L'Hôpital's rule tells us to take a derivative of the numerator and a derivative of the denominator. If you recall what we learned about derivatives in Chapter 8, the derivative of $\ln(x)$ with respect to[113] x gives us $\frac{1}{x}$ as the new numerator and the derivative of x with respect to x in the denominator gives us 1 as the new denominator. The new ratio is $\frac{1}{x}$ divided by 1, which is simply $\frac{1}{x}$ (because dividing by 1 doesn't have any effect). Since the $\frac{1}{x}$ approaches zero as x grows infinite, it follows that the sequence $\left\{\frac{\ln(n)}{n}\right\}$ converges to zero.

Before we discuss the convergence and divergence of series, we'll explore a couple of relevant formulas. The first formula is for **triangular numbers**. Recall that triangular numbers follow the pattern 1, 1+2, 1+2+3, 1+2+3+4, 1+2+3+4+5, 1+2+3+4+5+6, ... If we add the positive integers 1 thru n (including n), the result is equal to $\frac{n(n+1)}{2}$. For example, when $n = 5$, the triangular number is 1+2+3+4+5, for which the formula gives $\frac{5(5+1)}{2} = \frac{5(6)}{2} = \frac{30}{2} = 15$. If you add 1 thru 5, you'll see that the sum is indeed 15. As another example, when n = 10, the formula $\frac{10(10+1)}{2} = \frac{10(11)}{2} = \frac{110}{2} = 55$ gives the sum of 1 thru 10 (which you can verify with a calculator).

Another important formula is the formula for a **geometric series**. For a geometric series with exactly M terms, the series has the form $a + ar + ar^2 + ar^3 + ar^4 + \cdots + ar^{M-1}$, where a and r are constants. The constant a equals the first term and the constant r is the ratio of any term to the previous term. Each term is r times the previous term. In summation notation, the series looks like the formula below. The formula for the sum of this series is $\frac{a-br}{1-r}$, provided that r doesn't equal one, where b = ar^{M-1} is the last term in the series. We'll illustrate this formula with an example in the next paragraph.

---

[113] Formal calculus analyzes the ratio $\frac{\ln(x)}{x}$ in the limit that x approaches infinity and uses this to draw conclusions about $\frac{\ln(n)}{n}$ in the limit that n approaches infinity. The distinction is that x is a real variable that can be a decimal, whereas n is an integer.

$$\sum_{n=1}^{M} ar^n = ar + ar^2 + ar^3 + \cdots + ar^{M-1}$$

$$= \frac{a - br}{1 - r} = \frac{a - ar^M}{1 - r} = \frac{a(1 - r^M)}{1 - r}$$

Note: $br = (ar^{M-1})r = ar^{M-1+1} = ar^M$

For example, consider the geometric series $3 + 3(2) + 3(2)^2 + 3(2)^3 + 3(2)^4 + 3(2)^5$. You can count that this series has exactly 6 terms. Note that $M - 1 = 6 - 1 = 5$ for this series, and that the last term, $3(2)^5$, has an exponent of 5. The series simplifies to $3 + 6 + 12 + 24 + 48 + 96$. Each term is 2 times the previous term. The first term is $a = 3$, the ratio of consecutive terms is $r = 2$, and the last term is $b = 3(2)^5 = 3(32) = 96$. The formula tells us that the sum is $\frac{a-br}{1-r} = \frac{3-96(2)}{1-2} = \frac{3-192}{-1} = \frac{-189}{-1} = 189$. You can use a calculator to verify that $3 + 6 + 12 + 24 + 48 + 96 = 189$.

In calculus, we're mostly interested in **infinite** series, meaning series with an infinite number of terms. We're primarily interested in whether a given series converges or diverges, and there are a variety of tests to try to determine this.

If an infinite series **converges**, this means that the sum of an infinite number of terms is a finite value. Put another way, as you add more and more terms to the sum, the sum approaches a finite value. On the other hand, if an infinite series **diverges**, this means that the sum doesn't approach a finite value. Usually, this means that the sum grows infinite.

How can you add an infinite number of values together? Well, you can't. Yet calculus offers a way to figure out whether an infinite series has a finite or infinite sum. The 'trick' is to make a series of partial sums. For example, consider the series $1 + \frac{1}{4} + \frac{1}{9} + \frac{1}{16} + \frac{1}{25} + \frac{1}{36} + \frac{1}{49} + \frac{1}{64} + \cdots$ You can't add an infinite number of these terms together. But you can analyze the sequence (not series!) $1, 1 + \frac{1}{4}, 1 + \frac{1}{4} + \frac{1}{9}, 1 + \frac{1}{4} + \frac{1}{9} + \frac{1}{16}, 1 + \frac{1}{4} + \frac{1}{9} + \frac{1}{16} + \frac{1}{25}, \cdots$ This sequence is called the **sequence of partial sums** because each term is just part of the full sum. If the sequence of partial sums converges to a finite limiting value, the series converges. If instead the sequence of partial sums diverges, the series diverges.

You might be wondering, "How can you add an infinite number of values together and get a finite sum?" This is a good and important question. It turns out that an infinite number of values can make a finite sum. We'll illustrate this with an example.

Suppose that an inventor designs a robot that always travels halfway to its destination. The robot and a coin are placed one mile apart with level ground in between. The coin is programmed to be the robot's destination. The inventor presses a Go button on the robot. The robot travels exactly 1/2 a mile and stops. The inventor presses the Go button again. Now the robot travels 1/4 of a mile and stops. Next time the robot travels 1/8 of a mile, then 1/16 of a mile, then 1/32 of a mile, and so on. Each time, the robot travels halfway to the coin from its previous position. Obviously, the robot will never quite get there,[114] but the robot can get within an infinitesimal distance (which is just the tiniest positive distance) of the coin after a very large number of attempts. In the limit that the number of attempts grows infinite, the total distance traveled by the robot approaches one mile (without quite ever getting there). This shows that $\frac{1}{2} + \frac{1}{4} + \frac{1}{8} + \frac{1}{16} + \frac{1}{32} + \frac{1}{64} + \frac{1}{128} + \frac{1}{256} + \frac{1}{512} + \frac{1}{1024} + \cdots$ converges to 1. If you add these first ten terms with a calculator, you'll get approximately 0.9990234; if you add billions of terms, they will be extremely close to one; as the number of terms goes to infinity, the sum approaches one.

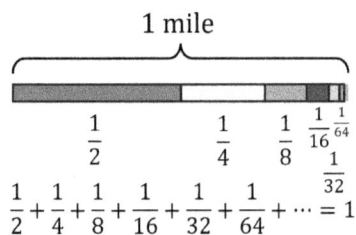

$$\frac{1}{2} + \frac{1}{4} + \frac{1}{8} + \frac{1}{16} + \frac{1}{32} + \frac{1}{64} + \cdots = 1$$

We can demonstrate that this series converges using the concept of the **sequence of partial sums** discussed previously. The sequence of partial sums is $\frac{1}{2}, \frac{1}{2} + \frac{1}{4}, \frac{1}{2} + \frac{1}{4} + \frac{1}{8}$,

---

[114] This bears similarity to one of Zeno's paradoxes. Zeno is one of the ancient Greeks who discovered challenges thinking about mathematical or physical concepts involving infinity. Perhaps Zeno would have reasoned that the robot would travel a total distance of very nearly one mile, but the thinking along one of Zeno's paradoxes would suggest that the robot will never reach the coin. Mathematically, Zeno has a point. Even if the robot continues this literally forever, the robot will still be an infinitesimal distance short of the goal. If you try to reach a destination by always going exactly halfway, you'd never get there. (Well, on a practical level, once you're less than an inch of the destination, you might feel you're close enough to say that you are indeed 'there.') Zeno might (well, it will be hard to ask him to confirm this; we can merely try to judge by his writings and the way he formulated his paradoxes) have said that if you always go halfway there at each step, it would take an infinite amount of time to reach your destination. (That's if you want to get all the way there and won't be content with being a fraction of an inch short of it.)

$\frac{1}{2}+\frac{1}{4}+\frac{1}{8}+\frac{1}{16},\frac{1}{2}+\frac{1}{4}+\frac{1}{8}+\frac{1}{16}+\frac{1}{32}$, ... This simplifies to $\frac{1}{2},\frac{3}{4},\frac{7}{8},\frac{15}{16},\frac{31}{32}$, ... The nth term of this sequence is $\frac{2^n-1}{2^n}$. For example, the 5th term is $\frac{2^5-1}{2^5}=\frac{32-1}{32}=\frac{31}{32}$. This sequence of partial sums converges to the limiting value of one. The numerator is always one less than the denominator, so, for example, when the denominator is 1,073,741,824 (corresponding to $n=30$), the ratio is $\frac{1,073,741,823}{1,073,741,824}$, which is very nearly one. When we can prove that the sequence of partial sums converges to a finite limiting value, as is the case in this example, we can prove that an infinite series converges.

Let's consider a few infinite series that are particularly important in series calculus. We'll begin with the infinite **geometric series**. Recall that a finite geometric series with M terms has a sum equal to $\frac{a-br}{1-r}$, where a is the first term, r is the ratio of any term to the previous term, and $b=ar^{M-1}$ is the last term. For an infinite geometric series, we just need to consider what happens to the last term, ar^{M-1}, in the limit that M grows infinite. If r < 1, then the last term approaches zero as M goes to infinity. For example, if r = 0.9, when you raise 0.9 to a very larger power, the number is very small. You can see this with 0.9^{100} which is approximately 0.000026561 and 0.9^{1000} which has more than forty zeroes between the decimal point and the first nonzero digit. Since the last term, b, approaches zero as M goes to infinity, the ratio $\frac{a-br}{1-r}$ approaches $\frac{a}{1-r}$ in this case. When r < 1, an infinite geometric series **converges** to $\frac{a}{1-r}$. Otherwise, an infinite geometric series **diverges**.

For example, the infinite geometric series $\frac{1}{2}+\frac{1}{4}+\frac{1}{8}+\frac{1}{16}+\frac{1}{32}+\frac{1}{64}+\frac{1}{128}+\frac{1}{256}+\frac{1}{512}+\frac{1}{1024}+\cdots$ converges to one. (This should sound familiar. We considered this same series two paragraphs ago.) For this series, the first term is a = $\frac{1}{2}$ = 0.5 and the ratio r = $\frac{1}{2}$ = 0.5 (since any term is one-half of the previous term). The series converges because r = 0.5 is less than one, and the formula tells us that the series converges to $\frac{a}{1-r}=\frac{0.5}{1-0.5}=\frac{0.5}{0.5}=1$.

As another example, the infinite geometric series 2 + 4 + 8 + 16 + 32 + 64 + 128 + 256 + 512 + 1024 + ⋯ diverges. In this case, the first term is a = 2 and the ratio r = 2 (since any term is twice the previous term). Since r = 2 isn't less than one, this series diverges.

Another important infinite series is the **harmonic** series $1+\frac{1}{2}+\frac{1}{3}+\frac{1}{4}+\frac{1}{5}+\frac{1}{6}+\frac{1}{7}+\frac{1}{8}+\frac{1}{9}+\frac{1}{10}+\cdots$ It turns out that the harmonic series diverges. The terms of the harmonic series use the formula $\frac{1}{n}$. The reason that the harmonic series is very important among infinite series is that it plays a critical role in the comparison test, which we will describe shortly.

If we make every other term of the harmonic series negative, we get an **alternating harmonic** series. Interestingly, when the terms of the harmonic series alternate signs, as in $1 - \frac{1}{2} + \frac{1}{3} - \frac{1}{4} + \frac{1}{5} - \frac{1}{6} + \frac{1}{7} - \frac{1}{8} + \frac{1}{9} - \frac{1}{10} + \cdots$, the series converges. (Technically, it converges conditionally, not absolutely, but this parenthetical note will only make sense to those who have learned series calculus.) But when every term has the same sign (in which case it isn't an alternating series), the series diverges. The alternating version of the series converges (conditionally) according to Leibniz's theorem, which is mentioned below.

Students who study series calculus learn a variety of tests that help to determine whether an infinite series converges or diverges. Following is a brief description of some of the tests that series calculus students learn how to apply. These are called **convergence tests** (except for the first one, which is instead called a divergence test).

- According to the **divergence test**, the series $a_1 + a_2 + a_3 + \cdots a_n + \cdots$ diverges if the nth term, a_n, doesn't approach zero in the limit that n grows infinite. (But beware: If a_n does approach zero, it doesn't guarantee that the series will converge. It could converge or diverge. More information is needed in that case. That's why this is a divergence test and not a convergence test.) For example, the series $\frac{2}{3} + \frac{3}{4} + \frac{4}{5} + \frac{5}{6} + \frac{6}{7} + \frac{7}{8} + \cdots$ diverges because the nth term, which is $a_n = \frac{n+1}{n+2}$, doesn't approach zero as n goes to infinity; it approaches one as n goes to infinity.

- The **comparison test** can be used to test for convergence or divergence. For this test, you need to already be familiar with some infinite series that you know converge or diverge. For example, you should know that an infinite geometric series of the form ar^{n-1} converges if $r < 1$ and diverges otherwise. Calculus students also know that an infinite series of the form $\frac{1}{n^p}$ converges if $p > 1$ and diverges otherwise. When $p = 1$, the series is the harmonic series $1 + \frac{1}{2} + \frac{1}{3} + \frac{1}{4} + \frac{1}{5} + \frac{1}{6} + \cdots$, which diverges. If $p < 1$, the series also diverges. If $p > 1$, the series converges. For example, $1 + \frac{1}{4} + \frac{1}{9} + \frac{1}{16} + \frac{1}{25} + \frac{1}{36} + \cdots$ converges because this is the case $p = 2$. (For example, $\frac{1}{5^2} = \frac{1}{25}$ and $\frac{1}{6^2} = \frac{1}{36}$.) Students learn how to compare series that they know to new series in order to determine whether the new series converges or diverges.

- The **integral test** applies to positive nonincreasing series. This means that the formula for a_n is a nonincreasing function of n. That is, if n increases by one, the next term is either smaller than the previous term or identical to the previous term. For example, $a_n = \dfrac{1}{n}$ for the harmonic series, and this is a nonincreasing function of n; when n increases by one, the fraction $\dfrac{1}{n}$ gets smaller. For the integral test, we make a function out of a_n in terms of the real variable x (which isn't necessarily an integer). For example, with the harmonic series where $a_n = \dfrac{1}{n}$, we would work with the function $f(x) = \dfrac{1}{x}$, replacing n with x. If the integral of $f(x)\,dx$ from 1 to infinity converges, so does the corresponding infinite sum, and if the integral diverges, so does the sum.

- The **ratio test** examines the ratio of two consecutive terms, $\dfrac{a_{n+1}}{a_n}$, in the limit that n goes to infinity. If the ratio is less than one in this limit, the series converges. If the ratio is greater than one in this limit, the series diverges. But if the ratio goes to one in this limit, a different test is needed. The ratio test is especially helpful for factorials. An exclamation mark is used to indicate a factorial, and it means to multiply successively smaller integers until you reach one. For example, the quantity 5! (read as "five factorial") means 5 times 4 times 3 times 2 times 1, and equals 120. The series $1 + \dfrac{1}{2} + \dfrac{1}{6} + \dfrac{1}{24} + \dfrac{1}{120} + \dfrac{1}{720} + \cdots$ can be expressed as $\dfrac{1}{1!} + \dfrac{1}{2!} + \dfrac{1}{3!} + \dfrac{1}{4!} + \dfrac{1}{5!} + \dfrac{1}{6!} + \cdots$ Each term has the form $\dfrac{1}{n!}$. The $(n+1)^{\text{th}}$ term is $\dfrac{1}{(n+1)!}$. The ratio $\dfrac{a_{n+1}}{a_n}$ equals $\dfrac{1}{(n+1)!}$ divided by $\dfrac{1}{n!}$, which simplifies to $\dfrac{n!}{(n+1)!}$ because the way to divide by a fraction is to multiply by its reciprocal. Now factorials have the neat feature that $(n+1)! = (n+1)$ times $n!$, since $(n+1)!$ means to multiply $n+1$ times n times $n-1$ times $n-2$ and so on until you reach one. Using this information, we find that $\dfrac{n!}{(n+1)!}$ simply equals $\dfrac{1}{n}$. In the limit that n approaches infinity, $\dfrac{1}{n}$ approaches zero. Since $\dfrac{n!}{(n+1)!}$ approaches a value less than one (since zero is less than one), this series converges.

- The **root test** examines a_n to the power of $\dfrac{1}{n}$ in the limit that n goes to infinity. If the ratio is less than one in this limit, the series converges. If the ratio is greater than one in this limit, the series diverges. But if the ratio goes to one in this limit, a different test is needed.

- **Leibniz's theorem** applies to alternating series if the sequence of absolute values is nonincreasing (like $\dfrac{1}{2} - \dfrac{1}{3} + \dfrac{1}{4} - \dfrac{1}{5} + \cdots$, where the fractions $\dfrac{1}{2}, \dfrac{1}{3}, \dfrac{1}{4}, \dfrac{1}{5}$, etc. are getting smaller) and where the sequence (not series!) of absolute values converges to zero. When these two conditions are both met, the alternating series converges (condi-tionally) according to Leibniz's theorem. This is one reason we know that the alternating harmonic series, $1 - \frac{1}{2} + \frac{1}{3} - \frac{1}{4} + \frac{1}{5} - \frac{1}{6} + \frac{1}{7} - \frac{1}{8} + \cdots$ converges (condi-tionally), even though the non-alternating harmonic series $\frac{1}{2} + \frac{1}{3} + \frac{1}{4} + \frac{1}{5} + \frac{1}{6} + \frac{1}{7} + \frac{1}{8} + \cdots$ diverges. (If the series of absolute values converges, the corresponding alternating series is said to converge absolutely. If the alternating series converges but the corresponding series of absolute values diverges, the alternating series is said to converge conditionally.)

Another kind of series that students learn about in series calculus is called a power series. A **power series** has the form $a_0 + a_1 x + a_2 x^2 + a_3 x^3 + a_4 x^4 + a_5 x^5 + \cdots$ This is unlike other series that we've mentioned in that it includes a variable (x). In a power series, each term has the variable raised to a higher power than the previous term. The constants a_0, a_1, a_2, etc. are called coefficients. An example of a power series is $1 + \frac{x}{2} + \frac{x^2}{4} + \frac{x^3}{8} + \frac{x^4}{16} + \frac{x^5}{32} + \cdots$ in this example, the coefficients are $a_0 = 1$, $a_1 = \frac{1}{2}$, $a_2 = \frac{1}{4}$, $a_3 = \frac{1}{8}$, etc. The coefficients have the formula $\frac{1}{2^n}$. For example, for $n = 3$, we get $a_3 = \frac{1}{2^3} = \frac{1}{8}$. The terms of this series have the form $\frac{x^n}{2^n}$.

Power series are useful in calculus because of the properties of **Taylor series** and a simpler version of Taylor series called **Maclaurin series**. A Taylor series has terms proportional to $(x - a)^n$, whereas a Maclaurin series has terms proportional to x^n; the constant a is zero for a Maclaurin series, making it simpler. The Maclaurin series for a function $f(x)$ is found by taking successive derivatives of $f(x)$ with respect to x and evaluating them at $x = 0$. Each term of the Maclaurin series has the nth derivative evaluated at zero times x^n divided by $n!$ (Recall that we discussed this factorial nota-tion, $n!$, earlier in this chapter.) For values of x where the Maclaurin series converges, it turns out that the series converges to $f(x)$. Because of this, Maclaurin series (and, more generally, Taylor series) provide a method of approximating a function using polynomials.

As an example, the Maclaurin series for the sine function (recall Chapter 5), $\sin(x)$, is equal to $x - \frac{x^3}{6} + \frac{x^5}{120} - \frac{x^7}{5040} + \cdots$ This series converges for any finite value of x. Here, the angle x must be in radians (not degrees). Note that 180 degrees corresponds to π radians.

Memory Test (Ch. 19)

1. What's the distinction between a sequence and a series? Give an example of each.

20 Which topics are covered in higher-level calculus courses?

The topics that we've explored in this book thus far are typically covered in two or three semesters of a calculus course (except with a lot more detail, way more algebra, plenty of trigonometry, and a more formal and abstract presentation). But there is actually more calculus beyond this, including vector calculus and the calculus of variations. In this chapter, we'll briefly discuss some topics from higher-level calculus courses.

In **multivariate calculus** (which is the calculus of multiple variables), the functions have two or more arguments instead of just one. For example, the multivariate function $f(x,y)$ is a function of two variables (x and y), and the multivariate function $g(x,y,z)$ is a function of three variables (x, y, and z). Such functions work much like the single-variable functions that we learned about in Chapter 5, except that they depend on two or more variables. As an example, consider $f(x,y) = x^3 - 4y$. To evaluate $f(2,1)$, replace x with 2 and replace y with 1 to get $f(2,1) = 2^3 - 4(1) = 8 - 4 = 4$.

Multivariate calculus is important because real-world functions tend to depend on several variables, not just one variable. For example, the pressure of an ideal gas depends on both its volume and absolute temperature:[115] $P(V,T) = NRT/V$. For real-world problems like trying to optimize the profit for a business, trying to predict the weather next Thursday, or trying to land a rocket on the moon, the functions involve multiple variables.

How do you take a derivative of a function that has multiple variables? The simplest kind of derivative in this context is called a partial derivative. If $f(x,y)$ is a function of x and y, to find a **partial derivative** of f with respect to x, we treat y as if it were a constant, and to find a partial derivative of f with respect to y, we treat x as if it were a constant. Here, x and y are independent variables, whereas the function f depends on the values of both x and y. When taking a partial derivative of a function

---

[115] In chemistry, students learn the ideal gas law, PV = NRT, where P is pressure, V is volume, N is mole number, R is the ideal gas constant, and T is the temperature in Kelvin. If you divide by V on both sides, you get P = NRT/V. If the number of moles of the gas is constant, then N is a constant, and P depends on the two variables V and T.

with respect to an independent variable, we treat the other independent variables as if they were constants. We'll demonstrate this with an example.

Consider the function $f(x,y) = x^3y^4$. To find a partial derivative of f with respect to x, we treat y as if it were a constant. Recall from Chapter 8 that a polynomial term of the form ax^b has a derivative given by the formula abx^{b-1}. If we treat y as if it were a constant, then $f = x^3y^4$ has the form ax^b where a = y^4 and b = 3. Our formula tells us that a partial derivative of f with respect to x is equal to $abx^{b-1} = (y^4)(3)\,x^{3-1} = 3x^2y^4$. To find a partial derivative of f with respect to y, we treat x as if it were a constant. When we treat x as if it were a constant, then $f = x^3y^4$ has the form ay^b where a = x^3 and b = 4. Our formula tells us that a partial derivative of f with respect to y is equal to $aby^{b-1} = (x^3)(4)y^{4-1} = 4x^3y^3$. The equations below show you how these partial derivatives look in a multivariate calculus course. Note that the partial derivative symbol (∂) is curved so that it looks different from the d used in ordinary derivatives.

$$f(x,y) = x^3y^4$$

$$\frac{\partial f}{\partial x} = \frac{\partial}{\partial x}x^3y^4 = y^4\left(\frac{d}{dx}x^3\right) = 3x^2y^4$$

$$\frac{\partial f}{\partial y} = \frac{\partial}{\partial y}x^3y^4 = x^3\left(\frac{d}{dy}y^4\right) = 4x^3y^3$$

In multivariate calculus, integrals are also different. When the function in an integrand may involve two or three variables, students need to perform a **double integral** or a **triple integral**. An example of a double integral is shown below. The way to perform a double integral is to integrate over one variable at a time. First, we look at the limits of integration to see if any of the limits involves a variable. In the example below, look at the limits of the y-integration: from y = 0 to x^2. Since the variable x appears in the limits of the y integral, students would need to integrate over y before integrating over x. When students integrate over y, they treat the independent variable x as if it were a constant (the same underlying principle we learned regarding partial derivatives). We're not going to do this double integral; it's way beyond the scope of this book. But we provided it as an example of calculus topics that come beyond first-year calculus.

$$\int_{x=0}^{1}\int_{y=0}^{x^2} f(x,y)\,dydx$$

Double and triple integrals are used to find volumes of three-dimensional solids, surface areas of curved surfaces, the center of mass of an object (which is the balancing point), or the moment of inertia of a rigid body (which tells you how easy or hard it is to change its angular velocity), for example. Double and triple integrals are also common in vector calculus, which we'll consider next.

Vector calculus involves quantities called vectors. When measuring physical quantities, some include direction and some do not. A quantity that includes direction is called a **vector**, whereas a quantity that doesn't have direction is called a **scalar**. We say that a vector has both a magnitude and a direction, whereas a scalar only has a magnitude. The word **magnitude** is a fancy way of saying 'how much.' The magnitude is just the number with units. For example, the distinction between speed and velocity is that speed is a scalar and velocity is a vector. If we say that a ball is traveling 45 m/s and is headed southwest, we have given the ball's velocity. The velocity of the ball includes a magnitude (in this case, 45 m/s) and a direction (southwest). If we just say that the ball is traveling 45 m/s but don't indicate which way it is moving, we have only given its speed. Many quantities that can be measured turn out to be vectors, including force, electric field, magnetic field, momentum, angular velocity, torque, and acceleration. Some measurable quantities that are scalars include mass, charge, volume, density, and wavelength.

The calculus involving vectors is so rich and applicable that there is an entire subject devoted to it: vector calculus. It is widely used in electromagnetism because electric and magnetic fields are vectors. (We'll see a glimpse of this in the next chapter when we briefly visit Maxwell's equations.)

Students learn about three different kinds of derivatives in vector calculus: the gradient, divergence, and curl. Each of these derivatives involves partial derivatives. The gradient operator acts on a scalar function of multiple variables and the result of the gradient is a vector. The divergence operator acts on a vector function of multiple variables and the result of the divergence is a scalar. The curl operator acts on a vector function of multiple variables and the result of the divergence is a vector. The symbols for all three operators (the gradient, divergence, and curl) involve a symbol called the del operator, which looks like an upside-down triangle. The symbols for the three operators appear below. The symbol for the gradient of the scalar function $f(x,y,z)$ is shown on the left, the symbol for the divergence of the vector function $A(x,y,z)$ is shown in the middle, and the symbol for the curl of the vector function $A(x,y,z)$ is shown on the right. (The little arrow on top of the A indicates that it's a vector. Textbooks usually use boldface to indicate a vector, whereas students writing by hand draw an arrow over the letter to indicate a vector because it isn't easy to draw boldface by hand.) The symbols for the gradient, divergence, and curl appear below. These are just the symbols used to represent them, not the formulas for how to calculate these quantities. (Those formulas are way beyond the prerequisites for this book.)

$$\nabla f = \text{grad } f \quad , \quad \nabla \cdot \vec{A} = \text{div } \vec{A} \quad , \quad \nabla \times \vec{A} = \text{curl } \vec{A}$$

The gradient, divergence, and curl operators have important physical interpretations.

- The **gradient** of the scalar function f(x,y,z) gives both the rate and direction of the fastest increase of the function. You may have heard the term 'gradient' before; if so, you might associate the term 'gradient' with slope or steepness (which is the basic idea). On a topological map, curves indicate where elevation is constant. Such curves are closely spaced where the gradient is larger and are more spread out where the gradient is smaller; here, the gradient tells you where the terrain is steepest.

- The **divergence** of the vector function A(x,y,z) provides a measure of the net flux of field lines radiating outward from a point. For example, in physics a proton is a positive charge and electric field lines radiate outward from it (like the spokes of a bicycle wheel); the divergence is positive where the field lines radiate outward.

- The **curl** of the vector function A(x,y,z) provides a measure of the net circulation of field lines at a point. For example, a long straight wire carrying a current produces magnetic field lines that circulate around the wire; there is a curl associated with these circulating magnetic field lines.

In vector calculus, the integrands of integrals involve vector functions. The integrals can be single, double, or triple integrals. Two important integrals in vector calculus are known as the divergence theorem (or Gauss's law) and Stokes's theorem. Each is a generalization of another integral called Green's theorem. We briefly mention each of these to give you a brief glimpse into the nature of vector calculus, but the underlying math is way beyond the scope of this book, so we'll try to keep it quick and simple.

- **Green's theorem** involves a vector field F(x,y) lying in the xy plane. Green's theorem involves the line integral of the vector field F(x,y); a line integral is an integral along a closed curve; a closed curve is one that encloses an area, like an ellipse or a triangle. A line integral is a single integral. Green's theorem also involves the surface integral of the z-component of the curl of F(x,y); the surface of integration is bounded by the closed path of the line integral. The surface integral is a double integral with respect to x and y. Green's theorem states that the line integral is equal to the surface integral. An interesting feature is that a single integral (the line integral) is equated to a double integral (the surface integral); the connection between these two integrals is that the closed path of the line integral is the boundary of the surface. For example, if the closed path is a circle, the area within the circle is a surface bounded by the closed path. The symbol C represents the path, the symbol S represents the surface bounded by the path, and dA is a differential element of the area lying on the surface. Since force is a vector (it has both a magnitude and a direction), it has components along the x- and y-axes, denoted F_x and F_y. (The components can be found by applying trigonometry. Students learn this part in a first-year physics course.) The circle in the integral symbol on the left indicates that the path is closed (meaning that it encloses an area; a circle encloses an area, whereas a line segment does not). The integrals below may seem intimidating, but that's because you haven't yet learned about vector calculus (which is beyond the scope of this book). The equation below offers a brief glimpse into what vector calculus looks like.

$$\oint_C F_x\,dx + \oint_C F_y\,dy = \iint_S \left(\frac{\partial F_y}{\partial x} - \frac{\partial F_x}{\partial y}\right) dA = \iint_S \left(\nabla \times \vec{\mathbf{F}}\right)_z dA$$

• The **divergence theorem** (or Gauss's law) relates the net flux of field lines through a closed surface to the divergence of the field, where the volume is bounded by the surface. The divergence theorem equates the double integral over a surface to the triple integral over a volume, where the volume is bounded by the surface. For example, if the surface is a sphere, the double integral is over the surface area of the sphere, while the triple integral is over the volume within the sphere; the volume is bounded by the spherical surface. In the context of physics, the divergence theorem is known as **Gauss's law**, which states that the net flux of field lines passing through any closed surface is proportional to the net charge enclosed by the surface. So, for example, if an imaginary sphere enclosed five protons and three electrons, there is a net positive charge inside the sphere and there will be a net flux of field lines passing through the sphere (meaning that there will be more field lines exiting the sphere than entering it, on a relative basis). If there were five protons and five electrons within the sphere, the net charge enclosed by the sphere would be zero, and the same number of field lines would exit the sphere as enter it. In the equation below, S represents a surface that encloses a volume V. (For example, S could be the surface of a sphere and V could be the volume inside of it.) The circle in the integral symbols on the left indicates that the surface is closed (meaning that it encloses a volume; a sphere encloses a volume, whereas a coffee mug does not, since coffee would fall out of a mug if turned upside down). The double integral on the left represents the net flux of field lines through the closed surface and the triple integral on the right represents the net charge within the volume of the surface. The divergence of the field appears in the integrand of the triple integral.

$$\oiint_S \vec{F} \cdot d\vec{A} = \iiint_V \nabla \cdot \vec{F} \, dV$$

• **Stokes's theorem** generalizes Green's theorem to the case where the closed curve (that is bounded by a surface) doesn't necessarily lie within a plane, but curves through three-dimensional space. (It is sometimes written as Stokes' theorem instead of Stokes's theorem.) In the equation below, C represents a path that encloses an area S. (For example, C could be the circumference of a circle and S could be the region within the circle.) The circle in the integral symbol on the left indicates that the path is closed (meaning that it encloses an area; a circle encloses an area, whereas a line segment does not). The curl of the field appears in the integrand of the double integral. Stokes's theorem and the divergence theorem (in the form of Gauss's law) are fundamental to the theory of electromagnetism; we see these laws in Maxwell's equations. We'll see Maxwell's equations briefly in the next chapter.

$$\oint_C \vec{F} \cdot d\vec{s} = \iint_S \nabla \times \vec{F} \, dA$$

Another branch of calculus that is rich and applicable enough to be its own course is the calculus of variations. The underlying idea behind the **calculus of variations** is to consider the initial state of a system (call it i), the final state of a system (call it f), and all of the possible ways that the system can proceed from the initial state i to the final state f. There could be an object traveling from point i to point f, and we can be thinking about all of the different possible paths that the object could take to go from i to f. But it could also be more abstract. There could be a thermodynamic system at an initial high temperature which will eventually cool to a lower equilibrium temperature, and we can be considering all of the ways that the thermodynamic system can pass through various states to reach equilibrium. Of all the possible paths (which may be literal like a rolling ball or more abstract like a thermodynamic system), there is some path which optimizes some parameter. For example, for a rolling ball, the path of least distance is a straight line connecting point i to f, but the path of least time may be a curved path (which may be the case if the ball changes speed as it rolls; we'll see an example of this when we consider the brachistochrone problem). For a thermodynamic system, the path of interest might be one that maximizes the entropy (or statistical disorder) of the system for a fixed energy, or it might be one that minimizes the energy of the system (for a fixed entropy). For any kind of system where it may be desirable to optimize some parameters, the calculus of variations may be applicable. The calculus of variations is fundamental to classical mechanics and quantum mechanics in physics, for example. We'll look at a couple of concrete examples conceptually (that is, without getting into the rich mathematics involved in it).

First, consider the following lifeguard problem. In the diagram below, there is a lifeguard at point A in a tower on the beach, and a distressed swimmer at point B in the ocean. The lifeguard wishes to reach the swimmer in the shortest possible time. The lifeguard can run on the sand twice as fast as the lifeguard can swim in the ocean. Which path should the lifeguard take in order to reach the swimmer in the least possible time (assuming that the swimmer doesn't move while the lifeguard travels towards the swimmer)?

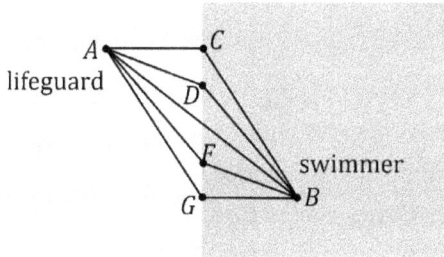

It turns out that the path of least distance, which is a straight line from A to B, isn't the same as the path of least time. Why not? If the lifeguard travels directly from A to B, the lifeguard will spend a considerable amount of time swimming. Since the lifeguard swims in water slower than the lifeguard runs on sand, the lifeguard can reach the swimmer in less time by taking a path that spends more time on sand and less time in water. But the path where the time spent in the water is a minimum, which is that path from A to G to B in the diagram above, isn't the path of least time either; although this path minimizes the time spent swimming, it adds considerably more distance to the path spent in the sand. The optimal solution lies somewhere between the path of least distance and the path with the least time spent in the water. The path from A to F to B is the path of least time in the diagram above. Along this path, the lifeguard spends a greater proportion of his time on the sand, where he runs faster than he swims, without adding too much distance to the trip. Using the calculus of variations, it's possible to solve for the exact angle with which the lifeguard should head in order to reach the swimmer in the least possible time.

It turns out that light does exactly the same thing as the lifeguard in the previous problem. When a ray of light travels at an angle from air to glass, for example, the ray of light **refracts**, meaning that it bends (it changes direction). Light travels much faster through air than it does through glass. According to **Fermat's theorem**, light travels along the path of least time. In the diagram below, a ray of light starts at point A in air and reaches point B in glass. The ray of light refracts (or bends) as shown below when it passes from air to glass. It turns out that this path is the path that minimizes the time it takes for light to travel from A to B. We can apply the calculus of variations to Fermat's theorem to find the angle that minimizes the time in this problem. The result is known as **Snell's law**, which is a formula for the angle of refraction.

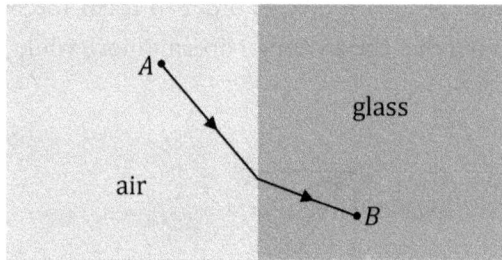

As another example of the calculus of variations, consider the brachistochrone problem. The **brachistochrone** refers to the path of least time in the following situation. There are two points: point A is higher than point B. A ring will slide along a wire starting at point A and finishing at point B. You have the freedom to shape the wire anyway that you wish, so long as the ring starts at A and finishes at B. Of the infinite possibilities, four representative paths (numbered 1, 2, 3, and 4) are shown below. Gravity pulls downward, accelerating the ring as it slides along the wire. The brachistochrone is the path for which the ring will reach point B from point A in the least amount of time (neglecting friction and air resistance). Looking at the diagram below, which path do you think would let the ring reach point B in the least amount of time? The answer may surprise you. (It's not the straight-line path, which is the path of least distance. Yet of the remaining three choices, the correct answer may still surprise you.) The calculus of variations can be used to solve for the path of least time, which is known as the **brachistochrone**, and it turns out to be curve 4 (the left curve) in the diagram below. This path is called a **cycloid** and has a fascinating feature. The cycloid actually takes the ring below point B; the ring falls below point B and then rises up at the end. The ring gains so much speed along the steep downward path at the beginning that it saves time compared to all other paths, despite the fact that the ring slows down along the upward path at the end. This can be demonstrated mathematically using the calculus of variations (or by conducting a careful experiment in a laboratory with minimal frictional effects).

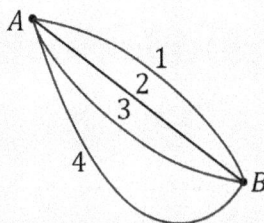

A related problem is the problem of a geodesic. If an object starts at point A and finishes at point B and is constrained to travel along the path of least distance, this path is called a **geodesic**. (If the object could travel through three-dimensional space the path of least distance would be a straight line, but if the object has to remain on a curved surface, the geodesic is often a curved path.) For example, if an object is constrained to lie on the surface of a sphere, the geodesic is part of the arc of a circle.

The mathematics behind the calculus of variations involves an integral known as the **action**. For example, in the Lagrangian formulation of classical mechanics, the integrand is the difference between the kinetic energy and potential energy, and the integral is over time. Optimization principles lead to the Euler-Lagrange equations for the motion of the object. The Lagrangian formulation, and a similar formulation called the Hamiltonian formulation, are two main methods that are widely used to solve a variety of physics problems.

What comes after calculus? There is always more math one can learn. Physicists, engineers, and mathematicians learn about **linear algebra** (which involves matrices and an important application called the eigenvalue problem) and **differential equations** (which are equations with applications in just about any field of math, science, or engineering that involve differential elements; these equations are solved by performing integrals). We won't try to list every course beyond this, but a few of the options include graph theory, group theory, topology, mathematical physics, and complex analysis.

Fill in the Blank (Ch. 20)

See if you can fill in each blank below.

1. A vector has a _____ and a _____.

2. A scalar has a _____, but doesn't have a _____.

21 What does calculus look like when it's applied?

When students first learn calculus, although it may appear scary at first, it's relatively straightforward. We start out with derivatives and integrals involving a single variable. In Chapters 8 and 14, we tried to make this look simple, but in an actual calculus course, it can get rather involved with a great deal of algebra and trigonometry. But in the real world, problems tend to involve multiple variables. Derivatives tend to involve partial derivatives (mentioned in the previous chapter). Integrals tend to involve double or triple integrals (mentioned in the previous chapter). Derivatives and integrals also tend to involve vectors (also mentioned in the previous chapter). With all of this, real world problems tend to look much more complicated than the single-variable derivatives and integrals that are taught in first-year calculus.

In this chapter, we'll take a brief look at some real calculus as it appears in the context of two topics from physics. This will offer a sense of what calculus really looks like when it's applied. Students spend years studying these subjects before they truly master the theory, and that's after they've already mastered first-year calculus. So, you shouldn't expect all the concepts from this chapter to make perfect sense the first time you read about them. Hopefully, you can recognize a few ideas that you've learned in this book, and observe that they look different than the simple derivatives and integrals that we've learned about in this book. If you learn how to paint and then visit a museum, you can appreciate the work of masters like Picasso or Rembrandt. This book is kind of like learning how to fingerpaint, and this chapter is like a museum where the artists are physicists like Maxwell or Schrödinger. Just try to appreciate their works of art.

It turns out that the effects of electricity and magnetism are intertwined. Electric fields and magnetic fields are different manifestations of a more general electromagnetic field. Light, which allows us to see, is itself an electromagnetic wave. The mathematics that describes electromagnetic fields can be condensed into four important equations, collectively known as **Maxwell's equations**. The integral form of Maxwell's equations are written below. The following bullet points briefly summarize the meaning of each equation. Half of these equations are forms of the divergence theorem (Gauss's law) and half are forms of Stokes's theorem, which we briefly encountered in the previous chapter.

- The top equation is **Gauss's law** in electricity, which states that the net flux of electric field lines through any closed surface is always proportional to the net electric charge enclosed by the surface. Gauss's law tells us about the distribution of electric charge in electrostatics (for example, it explains why more charge tends to accumulate where a surface has greater curvature), and helps us calculate electric fields when systems exhibit a great deal of symmetry (like a uniformly charged sphere). In this equation, E represents electric field, q represents electric charge (where "enc" stands for the charge enclosed within the volume V), and epsilon (the Greek letter ε_0) is a constant (the permittivity of free space; it's the reciprocal of 4π times Coulomb's constant, which is a proportionality constant in Coulomb's law for the attractive or repulsive force between two charged particles).

- The second equation is **Faraday's law**, which states that an emf is induced in a loop of wire if the magnetic flux through the area of the loop changes in time. Faraday's law explains how an electric generator can induce a current using a magnetic field, for example. Since a changing magnetic flux can result in electric current, Faraday's law provides a great illustration of how electric and magnetic fields are inherently intertwined; they are different manifestations of an electromagnetic field. In this equation, E represents electric field and B represents magnetic field.

- The third equation is **Gauss's law in magnetism**. It turns out that the net magnetic flux through any closed surface is always zero. We have never observed a magnetic monopole, which would be the magnetic equivalent of electric charge. You may be aware that magnets have north and south poles. If you split a magnet in half, you get two smaller magnets, each with a pair of north and south poles. No matter how many times you split the magnet in half, you'll never have just a north pole all by itself without a south pole.[116] This is due to Gauss's law in magnetism. (If anyone ever discovers a magnetic monopole, we'll have to revise this equation to account for it. There are physicists searching for them; the idea of a magnetic monopole is compelling in the sense that it would complete a sort of symmetry in Maxwell's equations. As it is, Gauss's law in magnetism and in

---

[116] Eventually, you'll just have a single atom left to split, and that's when things really get interesting.

electricity look quite different; the right-hand side is zero in magnetism.) In this equation, B represents magnetic field.

• The bottom equation is the generalized form of **Ampère's law**, which includes what is referred to as 'displacement current.' We use Ampère's law to calculate magnetic fields that are created by electric currents (where again you see a connection between electricity and magnetism). Just as a changing magnetic flux can induce an electric field (which causes emf and current to be induced in a loop of wire), a changing electric flux can induce a magnetic field (this is the part of the generalized form of Ampère's law that involves a displacement current). The previous sentence is part of the reason some physicists are searching for magnetic monopoles; they see the symmetry between electricity and magnetism already present in Maxwell's equations, and would like to see it completed in Gauss's laws. [117] In this equation, B represents magnetic field, E represents electric field, I represents current (where "enc" stands for the current enclosed in the path C), epsilon (ε_0) is the same constant from the top equation, and mu (another Greek letter, μ_0) is an analogous constant in magnetism (the permeability of free space, whereas epsilon is permittivity).

$$\oiint_S \vec{E} \cdot d\vec{A} = \iiint_V \nabla \cdot \vec{E}\, dV = \frac{q_{enc}}{\varepsilon_0}$$

$$\oint_C \vec{E} \cdot d\vec{s} = \iint_S \nabla \times \vec{E}\, dA = -\frac{\partial}{\partial t} \iint_S \vec{B} \cdot d\vec{A}$$

$$\oiint_S \vec{B} \cdot d\vec{A} = 0$$

$$\oint_C \vec{B} \cdot d\vec{s} = \iint_S \nabla \times \vec{B}\, dA = \mu_0 I_{enc} + \varepsilon_0 \mu_0 \frac{\partial}{\partial t} \iint_S \vec{E} \cdot d\vec{A}$$

---

[117] On the other hand, it's natural in some areas of physics for symmetry to be broken. For example, in the standard model of elementary particles, symmetry breaking is how the Higgs particle gives mass to particles like the electron. Without this symmetry-breaking mechanism, all particles would be massless. (At least, they wouldn't have rest-mass. All particles, even photons which have zero rest mass, have relativistic mass as a result of their motion.) This footnote is intended just for the benefit of those who know something about particle physics (or who might be motivated to try to learn something about it).

Maxwell's equations can alternatively be expressed in differential form. The differential form of these equations appear below. It may help to briefly review the divergence and curl from the previous chapter, as well as the divergence theorem.

• The top left equation involves the divergence of the electric field. This is the differential form of Gauss's law in electricity. When there is a net charge inside of a closed surface, there is a divergence of electric field lines (like the field lines that radiate out of a positive charge, like the spokes of a bicycle wheel). In this equation, E represents electric field, rho (the Greek letter ρ, which appears similar to a curvy p) is the charge density (charge per unit volume), and epsilon (ε_0) is the permittivity of free space.

• The top right equation involves the curl of the electric field. This is the differential form of Faraday's law. When the magnetic flux changes, the derivative on the right-hand side is nonzero, so the left-hand side must also be nonzero, which causes the induced electric field (which causes an emf and current to be induced in a loop of wire present in the changing magnetic flux). In this equation, E represents electric field, B represents magnetic field, and t is time.

• The bottom left equation involves the divergence of the magnetic field. Since the right-hand side is zero, magnetic field lines never diverge like the spokes of a bicycle wheel. Magnetic field lines tend to circulate, or at least travel along closed loops (not necessarily circles). This is the differential form of Gauss's law in magnetism. In this equation, B represents magnetic field.

• The bottom right equation involves the curl of the magnetic field. This is the differential form of Ampère's law. In this equation, B represents magnetic field, E represents electric field, J is the current density (current per unit area), epsilon (ε_0) is the permittivity of free space (a constant from electricity), and mu (μ_0) is the permeability of free space (a constant from magnetism).

$$\nabla \cdot \vec{E} = \frac{\rho}{\varepsilon_0} \quad , \quad \nabla \times \vec{E} = -\frac{\partial \vec{B}}{\partial t}$$

$$\nabla \cdot \vec{B} = 0 \quad , \quad \nabla \times \vec{B} = \mu_0 \vec{J} + \varepsilon_0 \mu_0 \frac{\partial \vec{E}}{\partial t}$$

When one-to-one interactions between charged particles and photons (for example) produce discernible effects, the phenomena are only explained by **quantum mechanics**, which exhibits much different behavior from classical mechanics. When a bowling ball travels along an alley and knocks over bowling pins, the principles of classical mechanics can help to predict and explain what happens. In this macroscopic situation, the individual interactions between particles and photons (or between quarks and gluons, etc.) don't produce discernible effects that are different from everyday human experience. If the effects of quantum mechanics did produce discernible effects in bowling, some fascinating outcomes would occur. Just imagine if the bowling ball was headed directly for the front pin, but wound up behind the pins without even touching any of them, or if the bowling ball were traveling straight down the center of an alley and suddenly wound up in an adjacent alley. Such outcomes seem ludicrous in the case of bowling, but in the realm of quantum mechanics, aren't so outrageous. If you shine an ultraviolet photon on a sheet of metal, for example, if the photon has a high enough frequency, it will carry enough energy to eject an electron from the metal (called a photoelectron). This is known as the **photoelectric effect**. It can be explained using the principles of quantum mechanics, but not by considering the principles of classical physics. If you shine a beam of ultraviolet light on a metal, the observed phenomena are governed by the quantum mechanics of the individual interactions between electrons and photons.

When we observe the interactions of particles on a one-to-one basis, the outcomes are governed by chance in the form of probability distributions. You can't predict the outcome of a single interaction, but if you repeat the process hundreds of times, you can predict the statistical distribution that will be formed. In quantum mechanics, the probabilities are determined by a quantity called the wave function. The wave function is represented by the Greek letter psi (ψ). We solve **Schrödinger's equation**, which is written below, in order to calculate psi, from which we can perform integrals to calculate probabilities or expectation values. For example, if an electron orbits a positively charged nucleus, we can solve Schrödinger's equation to find electron clouds, which are allowed orbitals where the electron might reside. It's a probability cloud; there is a probability of finding the electron in allowed regions called orbitals. As another example, if a particle travels in a region where the potential energy has a double well (like a double valley, like the letter W), but the particle's total energy (including kinetic energy) isn't enough to clear the top of the crest between the two wells, in classical physics it wouldn't be possible for the particle to wind up in the valley it didn't start out in, but in quantum mechanics there is a finite probability that the particle can 'tunnel' from one valley to the other. Schrödinger's equation helps us calculate this probability, and agrees with results that can be observed in a laboratory. This is an example of quantum mechanics that would seem like the bowling ball problem mentioned earlier if such effects were noticeable in everyday life. Schrödinger's equation is a second-order differential equation, which can be solved via integration (specifically, two different integrals and a lot of algebra are needed). The one-dimensional time-independent version of Schrödinger's equation appears below. (There is also a time-dependent and three-dimensional version of the equation, involving partial derivatives.) In the equation below, h-bar (which has a line through it) is Planck's constant over 2π (where Planck's constant is related to quantization of energy; that is, energy comes in discrete packets; for example, a single packet of light energy is called a photon, and its energy is Planck's constant times frequency), m is the mass of the particle, the Greek letter psi (ψ) is the wave function, V is the potential energy, and E is the total energy.

$$-\frac{\hbar^2}{2m}\frac{d^2\psi(x)}{dx^2} + V(x)\psi(x) = E\psi(x)$$

Did you learn the two main ideas from this book?

We said that Chapters 8 and 14 were arguably the two most important chapters. Let's see if you remember the main idea from each chapter. (If not, it isn't too late. This quiz is open book.)

1. On a graph of f(x), what does the derivative df/dx represent?

2. On a graph of f(x), what does the definite integral of f(x)dx from A to B represent?

Who is the author?

Dr. Chris McMullen has over 20 years of experience teaching university physics in California, Oklahoma, Pennsylvania, and Louisiana. Dr. McMullen is also an author of math and science books. Whether in the classroom or as a writer, Dr. McMullen loves sharing knowledge and the art of motivating and engaging students.

The author earned his Ph.D. in phenomenological high-energy physics (particle physics) from Oklahoma State University in 2002. Originally from California, Chris McMullen earned his Master's degree from California State University, Northridge, where his thesis was in the field of electron spin resonance.

As a physics teacher, Dr. McMullen observed that many students lack fluency in essential math skills. In an effort to help students of all ages and levels become fluent in mathematics, he published a series of math workbooks on fractions, long division, word problems, algebra, geometry, trigonometry, logarithms, calculus, probability, differential equations, complex numbers, and more. Dr. McMullen has also published a variety of science books, including astronomy, chemistry, and physics.

Would you like to report a possible typo?

We strive to perfect our books, but everyone is human. If you believe that you found a typo, visit the author's blog (www.chrismcmullen.com) and look for the Contact Me option.

Index

Are you looking for other math/science books for a general audience?

The author, Chris McMullen, Ph.D., also has accessible guides to astronomy and chemistry, a fascinating visual introduction to the geometry of the fourth dimension, and an introduction to Roman numerals. Another book, on the four-color theorem and basic graph theory, is also accessible, meaning that you don't need a background in mathematics in order to understand the main ideas.

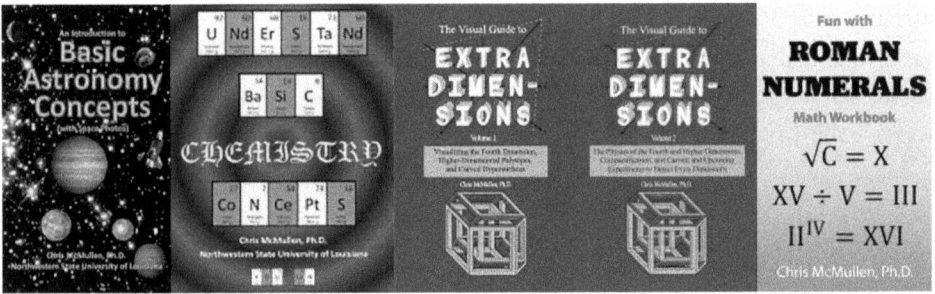

Would you like to try some real calculus?

As a review or supplement, what many students appreciate about Essential Calculus Skills Practice Workbook with Full Solutions is that the instructions are concise, the examples are clear and practical, and the full solution with explanatory notes is given for every exercise.

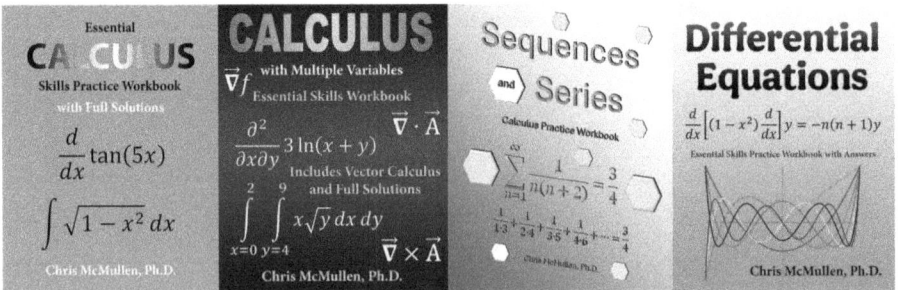

Do you enjoy puzzles?

If you like math puzzles, you may appreciate 300+ Mathematical Pattern Puzzles, which involves number pattern recognition and reasoning. If you want a greater challenge, try Pyramid Math Puzzle Challenge, which adds another dimension to the puzzles.

If you like anagrams (word scrambles), you might appreciate VErBAl ReAcTiONS, which features chemical words like GeNiUS (which are formed by stringing together symbols of elements from chemistry's periodic table to form words; here, Ge + Ni + U + S = GeNiUS). Don't worry, you don't need to know any chemistry to solve these puzzles; they're just word puzzles disguised as chemistry.

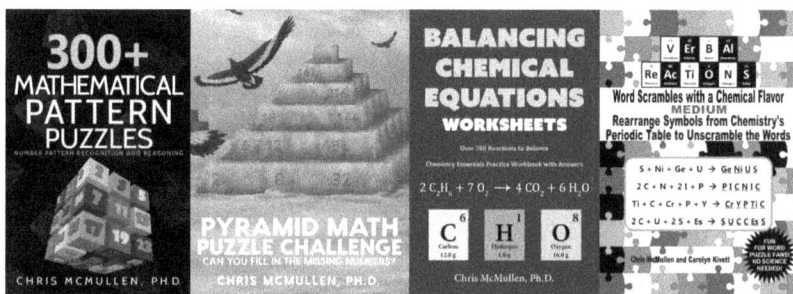

Would you like to improve your algebra fluency?

The author's most introductory and comprehensive algebra book, Master Essential Algebra Skills Practice Workbook with Answers, is great for students who are learning algebra. Students who know (or used to know) algebra and would like to review the main skills often prefer Algebra Essentials Practice Workbook with Answers.

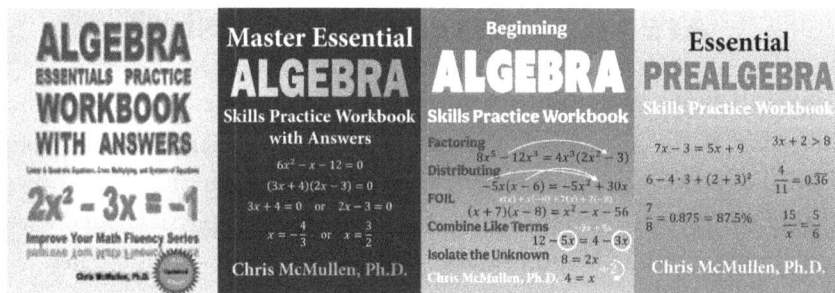

Are you curious about geometry?

Geometry Proofs Essential Practice Problems includes a variety of exercises to help you think about a variety of geometric properties. For a comprehensive introduction to geometry, try Plane Geometry Practice Workbook with Answers, Volumes 1-2.

The Four-Color Theorem and Basic Graph Theory isn't actually a geometry book, but its presentation is highly visual. You don't need a background in mathematics to understand the ideas in this fascinating tour of the four-color theorem.

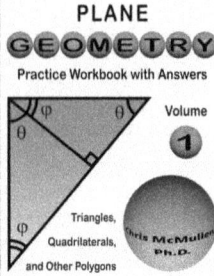

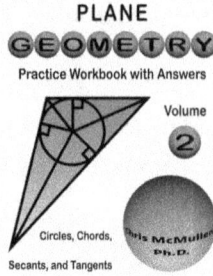

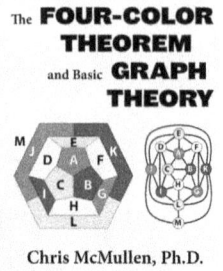

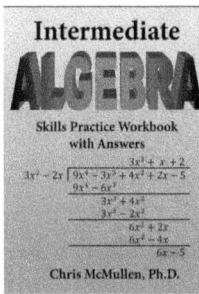

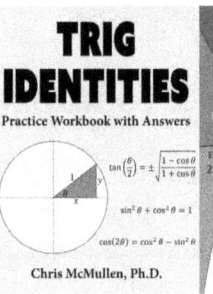

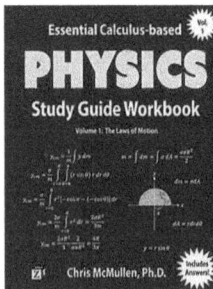

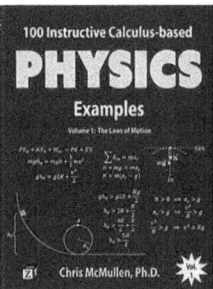

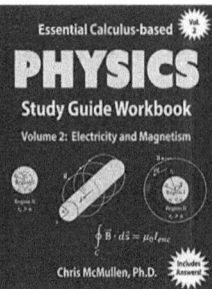